JOHN

ESSENTIAL READINGS

Gerald Suster
was educated at Highgate School and
Trinity Hall, Cambridge. A former
Head of History and Political Studies
at Albany College, he has been a
professional author, with a particular
interest in magic and mysticism, since
1979. He lives in London.

JOHN DEE

ESSENTIAL READINGS

Edited and Introduced by

GERALD SUSTER

First published 1986

Editorial matter
© **Gerald Suster 1986**

This work is dedicated to Francis
Israel Regardie, 1907-85,
a friend who kept the flame
of Dee alive.

British Library
Cataloguing in Publication Data

Suster, Gerald
John Dee: essential readings
1. Science — Early works to 1800
I. Title   II. Suster, Gerald
500   Q155
ISBN 0-85030-417-2

Crucible is an imprint of the
Aquarian Press, part of the
Thorsons Publishing Group.

Printed and bound in Great Britain

CONTENTS

CHRONOLOGY

1527   John Dee born in London.

1542–5 Studies at St John's College, Cambridge.

1546   Created one of the original Fellows of Trinity College, Cambridge.

1548–51 Studies at Louvain; visits Antwerp, and lectures on Euclid at Paris.

1551–3 Tutors Robert Dudley, later Earl of Leicester.

1553   Two church livings conferred on Dee by Edward VI.

1555   Imprisoned under Queen Mary on suspicion of casting enchantments against her.

1558   Fixes date for the coronation of Queen Elizabeth I by casting horoscope.

1563   Travels to Antwerp, Zurich, Rome, and Hungary.

1564   Writes *Monas hieroglyphica*.

1565   Marries Katherine Constable.

1570   Publication of H. Billingsley's first English translation of Euclid, with a celebrated Preface by Dee.

1571   Visits Lorraine.

1575   Marries for the second time.

1576   His (unknown) second wife dies.

1577   Publication of Dee's *General and Rare Memorials pertaining to the Perfect Art of Navigation*.

1578   Marries Jane Fromond. Visits Frankfurt-on-Oder.

1582   Dee's abortive attempt to introduce the Gregorian Calendar to England. Meets Edward Kelley.

1583–9   Dee, Kelley, and their wives travel in Europe, including periods of residence with the King of Poland and the Emperor Rudolf in Prague. Magic and alchemy are major preoccupations.

1589   Dee and his wife return to England.

1595   Death of Kelley in Prague.

1596   Queen Elizabeth makes Dee Warden of Manchester College.

1605   The Fellows of Christ's College force Dee to relinquish his post.

1608   Death of Dee in poverty and obscurity.

INTRODUCTION

Hee had a very faire cleare rosie complexion; a long beard as white as milke; he was tall and slender; a very handsome man . . . he wore a Gowne like an Artist's gowne, with hanging sleeves, and a slitt; a mighty good man he was.

John Aubrey, *Brief Lives*

UNTIL comparatively recently, John Dee was regarded as an isolated crank on the margin of Tudor history, beyond the pale of serious academic consideration, and of interest only to a small minority of antiquarians and occultists. Even today, the *Encyclopaedia Britannica* gives Dee just one small paragraph of scanty and under-researched information — a sad fate for a man who was revered in his time as the most learned man in all Europe.

Recent years, however, have seen a steady revaluation: the admirable scholarship of the late Dame Frances Yates, for instance, has clearly demonstrated Dee's central position in the history of sixteenth-century thought. The present work is an endeavour to introduce the general reader to this highly complex figure. All who have turned their attention to Dee acknowledge that more scholarship is sorely required. So it is: but a guide for the student who is approaching Dee studies for the first time is equally necessary.

John Dee was born in London on 13 July 1527, the son of Rowland Dee, a courtier in attendance on King Henry VIII. From 1542-5, he studied at St John's College, Cambridge, later writing of this period:

I was so vehemently bent to studie, that for those yeares I did inviolably keepe this order; only to sleepe four houres every night; to allow to meate and drink (and some refreshing after) two houres every day; and of the other eighteen houres all (except the time of going to and being at divine service) was spent in my studies and learning. [1]

In 1546, he became one of the original Fellows of Trinity College, Cambridge. That same year he constructed a flying machine for a performance of *Peace* by Aristophanes. Unfortunately, this ingenious feat formed the foundation of the accusation which would haunt him all his life: that he was a practitioner of the black arts and a conjurer of evil spirits.

From 1548–51, Dee continued his studies at Louvain, a university which received financial support from the Pope and the Emperor Charles V and which enjoyed an international reputation for Civil Law and Mathematics. He also visited Antwerp, and in a remarkable performance for a young man of twenty-three lectured successfully on Euclid in Paris, 'a thing never done publiquely in any University in Christendome'.[2]

Returning to England, he spent 1551–3 as the tutor of Robert Dudley, son of the Lord Protector Northumberland and later Earl of Leicester; and in 1553 two Church livings, the rectories of Upton-on-Severn, Worcestershire and Long Leadenham, Lincolnshire, were conferred on him by Edward VI. However, the accession of Queen Mary caused an unpleasant reversal of Dee's fortunes, for in 1555 he was imprisoned on suspicion of casting enchantments against her. He was eventually released, though his cellmate, Barthlet Grene, was burned to death, and he endeavoured to recover his credit. It was Dee who cast the horoscope which fixed the date for the coronation of Queen Elizabeth I in 1558.

The ensuing years were busy, productive, and successful. Dee became known at court and enjoyed the acquaintance of Leicester, his former pupil, Sir Philip Sydney, Sir William Cecil, and many other notables, including — and especially — the Queen herself. He read, wrote, studied, and was regarded as a person of consequence, influence, and great learning. Explorers such as Frobisher and Sir Humphrey Gilbert consulted him on matters of navigation. He collected books for what became the largest and finest library in England. In 1563, he travelled to Antwerp, Zurich, Rome, and Hungary; and in the following year he wrote the celebrated *Monas hieroglyphica*.

Biographers differ as to whether Dee was married two or three times, and a number omit all mention of his first wife. However, if *Chancery Proceedings*, Series II, Bundle 49, No: 44 can be trusted, she was Katherine Constable, a grocer's widow from the City. It was perhaps at this time that Dee moved to a house at Mortlake, for it is known that he was established in the village before 1570.

That year saw the publication of the first English translation of Euclid by H. Billingsley, later Lord Mayor of London, to which Dee wrote a deservedly famous Preface. In 1571 he visited Lorraine. By 1575, Katherine Constable must have died, for Dee married again. It appears that nothing at all is known about his second wife except that she died in the following year.

Dee's *General and Rare Memorials pertaining to the Perfect Art of Navigation* was published in 1577; the following year he visited Frankfurt-on-Oder and was married for the third time, to Jane Fromond, a lady-in-waiting to Lady Howard of Effingham.

Dee's abortive attempt to introduce the Gregorian Calendar to England took place in 1582, the fateful year in which he also met the man with whom his name is so often associated — Edward Kelley. Many have wondered how it was possible for an intelligent man like Dee, skilled in classical studies, navigation, mathematics, logic, literature, and philosophy, to occupy himself with alchemy, magic, and the calling of spirits, with Kelley aiding and abetting him. This is a question to which we shall return. For the present, let it suffice that the occult sciences were Dee's primary concern from 1583 to 1589, when he travelled in Europe with Kelley and their respective wives, residing at times with the King of Poland and the Emperor Rudolf in Prague.

In 1589, Dee and Kelley parted, Dee returning to England with his wife. Kelley died in Prague in 1595. If Dee hoped for a triumphant return and a glorious reception in his native country, he was sorely disappointed. Offers of assistance and patronage failed repeatedly to materialize and he was increasingly tormented by financial problems and by slander: in his absence on the Continent a mob had broken into and plundered his house at Mortlake. Finally, in 1596, Queen Elizabeth made him Warden of Christ's College, Manchester.

His time there was not an especially happy one. The Fellows objected to the extensive scheme of reform which he attempted to carry through; the assaults on his reputation grew in vehemence after the accession of the witch-hating King James I in 1603; and in 1605, the Fellows of Christ's College forced Dee to relinquish his post. He returned to Mortlake a widower, Jane having died shortly before.

His last years were sad as he sank into obscurity and increasing opprobrium, 'very poor, enforced many times to sell some book or other to buy his dinner'.[3] He died in 1608.

'Here lived the learned Mr John Dee who was one of the ornaments of his age but was mistaken by the ignorant for a conjurer', wrote John Aubrey in *Perambulation of Surrey: Mortlake*.[4] But Aubrey was one of the few who paid homage to Dee's posthumous reputation, which now suffered the further indignity of ridicule. By 1659, Meric Casaubon[5] was writing of Dee as a mere fanatic deluded by devils, a view accepted by Dee's first biographer, Dr Thomas Smith,[6] who called his subject 'the sport, the laughing-stock and the prey of daemons'. By the nineteenth century, Dee had even been robbed of his integrity and was condemned as 'dead to all moral distinctions, and all sense of honour and self-respect'.[7] At best a gullible fool and at worst an unsuccessful charlatan, Dee was dismissed in the *Biographia Britannica* as 'extremely credulous, extravagantly vain and a most deluded enthusiast'.

Yet Dee's reputation stubbornly refused to expire altogether. His name continued to be respected by the few who study the occult sciences: his own son; Elias Ashmole; and in the nineteenth century, the Hermetic Order of the Golden Dawn, whose work on Dee was continued in the present century by Alesiter Crowley, Israel Regardie, and others. It is ironic that Dee's magical studies kept his name alive for over three centuries: for it was precisely this factor which led orthodox scholars and historians to ignore him as unworthy of concern.

Fortunately, 1909 witnessed the commencement of a less hysterical approach to Dee with the publication of Charlotte Fell Smith's biography.[8] In *Tudor Geography* (1930), E. G. R. Taylor clearly demonstrated Dee's importance in terms of English navigation and gave him 'an honoured place in the History of Geography':[9] and F. R. Johnson[10] re-established Dee's position as a major figure in the development of astronomy, mathematics, science, and Renaissance thought. The work of Frances Yates[11] investigated this and many other aspects of Dee's influence on Tudor England and the Continent, concluding that the world picture of the Renaissance was in fact that of John Dee: and a recent biographer, the late Peter French,[12] accepted this judgement and further demonstrated its truth. Although Professor Wayne Shumaker[13] has warned sternly against over-estimating Dee's significance, he has nevertheless found him significant enough to be worthy of a fine contribution to Dee studies: nor should one forget the enthusiasm of another recent biographer, Richard Deacon.[14]

One curious result of this reassessment has been the occasional appearance of Dee on film and in popular culture. For example, in Derek Jarman's film *Jubilee*, Dr Dee grants Queen Elizabeth a vision of London in the 1970s: and in an American comic recently brought to my attention, Dee is credited with writing the plays of Shakespeare and secretly founding the United States of America.

The claims made seriously on his behalf are impressive:

He was the most learned man in Europe.

He brought European thought, philosophy, and wisdom to England.

He coined the phrase 'The British Empire' and was a prime mover behind Elizabethan Imperialism.

He was a key figure in matters of navigation and exploration.

He was one of the leading mathematicians of his age.

He was adept in Neoplatonic, Hermetic, and Cabalistic philosophy.

He was a noted antiquarian.

His mechanical skills were highly respected by English mechanicians.

He was expert in astronomy and astrology.

He influenced the development of English poetry by his intercourse with the Sudney circle, and this reflected in Spenser's *The Faerie Queen*. [15]

In *The Tempest*, Shakespeare based Prospero on Dee. [16]

And, though many may find the fact curious, Dee's work on magic remains revered by practising occultists. [17]

We shall be looking at the nature of these claims: but even if they are only partly true, Dee remains a remarkable figure. In this anthology I have tried to select material from every period of Dee's eventful life. My presentation has been deliberately inconsistent, a policy which requires a word of explanation.

Since 'the style is the man', there was an inevitable temptation to let Dee speak for himself. In many instances, that is exactly what has been done, but sixteenth-century vocabulary, grammar, and syntax can weary the modern reader and interfere with appreciation. When possible, therefore, I have included adequate nineteenth-century transliterations of Dee's writing. Finally, there are selections which required twentieth-century transliteration if they were to be made accessible. Here I have been guided by an endeavour to render the spirit of the original without falling into the pit of banality which disgraces modern rescensions of the Bible or the Book of Common Paryer.

Each extract is accompanied by a note on the aspect of Dee under consideration and a few words on my reason for the choice and suggestions regarding the content.

1

A SUPPLICATION TO QUEEN MARY

FOR THE RECOVERY AND PRESERVATION OF ANCIENT WRITERS AND MONUMENTS [1556]

WHERE did Dee get his ideas from? Fortunately, the matter has been studied in depth by Frances Yates and Peter French, who rightly stressed the importance of examining the contents of Dee's library.

'We find that Dee was a man of universal interests: the range of books on his shelves was extraordinary,' writes French, who adds: 'If the essential requisite of a university is an excellent library, F. R. Johnson has pointed out that Dee's home at Mortlake might truly be considered the scientific academy of England during the first half of Elizabeth's reign.'[1] Dee compiled a catalogue on 6 September 1583 and claimed that it numbered 'in all neere 4,000: the fourth part of which were written bookes'.[2] This becomes all the more remarkable if we bear in mind the fact that in 1582, the Cambridge University Library possessed merely 451 books and manuscripts.[3]

In the sixteenth century it was still just possible for one man to have universal knowledge, and this appears to have been Dee's ambition. His library included the complete works of Plato and Aristotle; the dramas of Aeschylus, Euripides, and Sophocles, Seneca, Terence, and Plautus; and the writings of Thucydides, Herodotus, Homer, Ovid, Livy, and Plutarch. There were many works on religion and theology: the Bible *and* the Koran; St Thomas Aquinas *and* Luther and Calvin. Every major work by contemporary British antiquaries was present, in addition to every major work on science, mathematics, and geography. The place of mysticism and magic in the scheme of things was stressed rather than neglected, and represented by Plotinus, Roger Bacon, Lull, Albertus Magnus, Ficino, Pico della Mirandola, Paracelsus, Trithemius, and Agrippa. As Frances Yates remarked: 'The whole Renaissance is in this library.'[4]

Although he enjoyed periods of prosperity, Dee was never a wealthy man, yet he nevertheless lavished £3000 on acquiring his collection, roughly the equivalent of £110,000 in today's terms.[5] He panted after knowledge as a miser after gold, and endeavoured from intellectual diversity to fashion an harmonious wisdom.

However, for Dee it was not sufficient merely to possess knowledge for its own sake: it also had to be applied. It is therefore appropriate to begin with an extract containing a practical proposal for the extension of knowledge.

Dee's *Supplication to Queen Mary* urges, with all the arts of courtier which his age required, the creation of a Royal Library. Unfortunately, no action was taken: but some may discern the seeds here of the future British Library.

❧

TO THE QUEEN'S MOST EXCELLENT MAJESTY

IN MOST humble wise complaining, beseeches your Highness, your faithful and loving subject, John Dee gentleman, to have in remembrance how that, among the exceeding many most lamentable displeasures that have of late happened to this realm, through the subverting of religious houses[6] and the dissolution of other assemblies of godly and learned men, it has been, and forever among all learned students shall be judged not the least calamity, the spoiling and destruction of so many and so notable libraries, wherein lay the treasure of all Antiquity and the everlasting seeds of continual excellence within this your Grace's realm.

But, albeit that in those days many a precious jewel and ancient monument did utterly perish (as at Canterbury did that wonderful work of the sage and eloquent *Cicero de Republica*, and in many other places the like): yet if in time, great and speedy diligence be shown, the remnants of such incredible store, as well as theological writers and those in the other liberal sciences, might be saved and recovered: which now in your Grace's realm are being dispersed and scattered, yea; and many of them in unlearned men's hands do still yet (in this time of reconciliation) daily perish, and perchance on purpose by some envious person enclosed in walls, or buried in the ground, to the great injury of famous and worthy authors and the pitiful hindrance of the learned in this your Highness' realm: whose travails, watchings and pains might greatly be relieved and eased; because such doubts and points of learning as much cumber and vex their heads are most pithily in such old monuments debated and discussed.

Therefore your said suppliant makes most humble petition unto your Majesty that it might stand with your goodwill and pleasure for such order and means to take place as your said suppliant has devised for the recovery and continual preservation of all such worthy monuments as yet are extant, either in this your Grace's realm of England, or elsewhere in the most part of all Christendom.

Whereby your Highness shall have amost notable library, learning wonderfully be advanced, the passing excellent works of our forefathers from rot and worms be preserved, and also hereafter continually, the whole realm may (through your Grace's goodness) use and enjoy the whole incomparable treasure so preserved: for now, no one student, no, nor any one College, has half a dozen of those excellent jewels; but the whole stock and store thereof is drawing nigh to utter destruction and extinction, while here and there by private men's negligence (and sometimes malice), many a famous and excellent author's book is rent, burnt or suffered to rot and decay.

And your said suppliant is so much the more willing to move this suit unto your Highness, because by his said device, your Grace's said library might in very few years most plentifully be furnished, and that without any one penny charge unto your Majesty, or doing injury to any creature.

Finally, in the erecting of this your Royal Library, your Grace shall follow in the footsteps of all the famous and godly princes of old time, and also do like the worthy Governors of Christendom in those days: but far surmounting them all, both in the store of rare monuments and likewise in the incredible fruit, which of this your Highness' act will follow before long. The merit whereof shall rebound to your Majesty's honourable and everlasting fame here on earth, and undoubtedly in heaven highly be rewarded: as knoweth God, Whom your said suppliant most heartily beseeches long to preserve your Grace in all prosperity. Amen.

★   ★   ★

ARTICLES CONCERNING THE RECOVERY AND PRESERVATION OF THE ANCIENT MONUMENTS AND OLD EXCELLENT WRITERS: AND ALSO CONERNING THE ERECTING OF A LIBRARY WITHOUT ANY CHARGES TO THE QUEEN'S MAJESTY, OR DOING INJURY TO ANY OF THE QUEEN'S SUBJECTS, ACCORDING TO THE TENOR AND INTENT OF A SUPPLICATION TO THE QUEEN'S GRACE IN THIS BEHALF EXHIBITED BY JOHN DEE, GENTLEMAN. A. 1556, THE XV DAY OF JANUARY.

1. *Imprimis*, the Queen's Majesty's commission to be granted for the

seeing and perusing of all places within this her Grace's realm where any notable or excellent monument may be found, or is known to be. And the said monument or monuments so found and had by the said Commissioner then; of the former possessor in the Queen's Majesty's name to be borrowed, and so nevertheless to be restorable to the said former possessor after such convenient time, wherein of every such monument one fair copy may be written, if the said former possessor be disposed to have the said monument or monuments again; and thereupon either he or his assignees do at the said Library (the place whereof is by the Queen's grace to be appointed) demand the said monument or monuments by bill assigned with the hand of the said Commissioner (wherein both the name or names of the said monument or monuments is or are particularly expressed) and also the convenient time for the said restitution prescribed.

2.   That it may be referred to my Lord Cardinal's Grace and the next Synod to conclude an order for the allowance of all necessary charges, as well toward the riding and journeying for the recovery of the said worthy monuments; as also for the copying out of the same, and framing of necessary stalls, desks and presses, fit for the preservation and use of the said monuments in the Queen's Majesty Library aforesaid.

3.   That the said Commission be with speed dispatched for three causes especially: first, lest after this motion is made, the spreading of it abroad might cause many to hide and convey their good and ancient writers (which nevertheless were very ungodly done, and a certain token that such are not sincere lovers of good learning). Secondly, that by the travail of these three months, February, March and April next going before the Synod, in May next appointed, the said Synod may have good proof whereby to conjecture how this matter will take success. And thirdly, upon the said trial of three months, the proportion of the charges in riding and writing may the better be weighed, what they will in manner amount to.

4.   A fit place to be forthwith appointed for the said monuments to be sent to, until the said Library may be made apt in all points necessary; and that in this said place, before or at the Synod-time, the said monuments may be viewed and perused according to the pleasure of my Lord Cardinal's Grace and the said next Synod.

5.   Finally, that by further device of your said suppliant, John Dee (God granting him life and health) all the famous and worthy

monuments that are in the most notable Libraries beyond the sea (as in the Vatican at Rome, San Marco at Venice, and the like at Bononia, Florence, Vienna &c.) shall be procured unto the said Library of our sovereign Lady and Queen, the charges thereof (beside the journeying) to stand in the copying of them out, and the carriage into this realm only. And as concerning all other excellent authors printed, that they likewise shall be gotten in wonderful abundance, their carriage only into this realm to be chargeable.

[From *The Autobiographical Tracts of Dr John Dee*, ed. J. Crossley, 1851. Transliteration: GS.]

2

PROPAEDEUMATA APHORISTICA
[1558]

IF THE truth of a science and the skill of its practitioners are to be judged purely in terms of results, then there is something to be said for astrology and for Dee as an astrologer, for the horoscope he cast to determine the most propitious day for the coronation of Queen Elizabeth I was followed by 'the golden days of good Queen Bess'.

These days, of course, few educated people take astrology seriously, but in the sixteenth century, it was considered a mark of a learned man. Astronomy was then its mere lady-in-waiting. One required skill in astronomy to ascertain the positions of the planets and stars, but skill in astrology to ascertain their meaning.

'O comfortable allurement,' Dee rhapsodized, 'O ravishing persuasion, to deale with a Science whose subject is so Ancient, so pure, so excellent, so surmounting all creatures, so used of the Almighty and incomprehensible wisdom of the Creator in the distinct creation of all creatures, in all their distinct parts, properties, natures and virtues, by order and most absolute number, brought from Nothing to the Formalities of their being and state!'[1]

Clearly, when Dee refers to 'order and absolute number' he is thinking of astrology as a science somewhat more profound than the horoscopes of our popular newspapers and he endeavoured to formulate his conception of the subject in *Propaedeumata Aphoristica* (1558), henceforth referred to as the *Aphorisms*. It is fortunate that a translation from the Latin has at last been done by Professor Wayne Shumaker, whose splendid contribution *John Dee on Astronomy*[2] also contains his commentary and a first-class introduction by J. L. Heilbron.

As Shumaker and Heilbron point out, Dee 'mathematically furnished up the whole method of Astrology'. Heilbron cautions against over-estimating Dee as a mathematician. For instance, Dee's name is not to be found in van Roomen's *Idea Mathematicae Pars Prima* (Antwerp, 1593), which lists the chief mathematicians of the later sixteenth century. It is allowed, however, that he was 'competent and knowledgeable'.

According to Heilbron, Dee perceived astrology as deriving from Number (as in the ancient aphorism 'God is the Grand Geometer') and as 'an inclusive, quantitative, physical science, assisted by experiment and aimed at both understanding and control of natural processes'. It is hardly a simple process, for in the *Aphorisms*, Dee directs the conscientious astrologer to discriminate between the influences of no less than 25,000 sorts of planetary configurations.

The *Aphorisms* are dedicated 'to the very distinguished Gentleman, Master Gerardus Mercator³ of Rupelmonde, Renowned Philosopher and Mathematician', and the Preface includes the following list of Dee's works so far:

1. Concerning Precision in Mathematics: a work of mathematical demonstration in sixteen books.
2. Concerning the Distances of Planets, Fixed Stars, and Clouds from the Center of the Earth, and Concerning the Discovery of the True Magnitudes of all the Stars: a demonstration in two books.
3. Of Burning Glasses: a demonstration in five books.
4. Of the Perspective Used by the Most Skilled and Famous Painters: a demonstration in two books.
5. Of the Third and Chief Part of Perspective, which Treats the Refraction of Rays: a demonstration in three books.
6. Of the Great Conveniences of the Celestial Globe: two books.
7. The Mirror of Unity, or Apology for the English Friar Roger Bacon; in which it is taught that he did nothing by the aid of demons but was a great philosopher and accomplished naturally and by ways permitted to a Christian man the great works which the unlearned crowd usually ascribes to the acts of demons: one book.
8. Concerning a New System of Navigation: two books.
9. Concerning Various Uses of the Astronomical Ring: one hundred chapters, one book.
10. Concerning a Subterranean Passage: one book.
11. Concerning the Triangle and the Analogical Compass: three books.

The main body of the text follows:

APHORISM I

As God created all things from nothing against the laws of reason and nature, so anything created can never be reduced to nothing unless this is done through the supernatural power of God and against the laws of reason and nature.

II

In actual truth, wonderful changes may be produced by us in natural things if we force nature artfully by means of the principles of pyronomia.[4] I call Nature whatever has been created.

III

Not only are those things to be said to exist which are plainly evident and known by their action in the natural order, but also those which, seminally present, as it were, in the hidden corners of nature, wise men can demonstrate to exist.

IIII

Whatever exists by action emits spherically upon the various parts of the universe rays which, in their own manner, fill the whole universe. Wherefore every place in the universe contains rays of all the things that have active existence.

[Note: the reader may discern in this idea of 'rays' similarities with the arguments of contemporary astrology. The essence of the matter is in Aphorism VII]

VII

The effects of any rays pouring from one thing upon diverse things are different.

[Note: the doctrine of correspondences, which is so succinctly put in Aphorism IX, is one of the major keys to the understanding of Renaissance Hermetic thought, which will be considered later.]

IX

Whatever is in the Universe possesses order, agreement and similar form with something else.

[Note: the above is also a clearer way of putting the Hermetic maxim 'As above, so below'. This idea is further expanded in Aphorism XXIII and Aphorism XXV, looking back at the idea of 'the music of the spheres' and perhaps forward to certain puzzling developments in the theory of contemporary quantum physics.]

XXIII

'That thoughts obey bodies and do not belong among insensible things, existing as they do through bodily perturbations' — what philosopher does not harp on this, and what mortal does not know it through almost daily experience? as also that 'The body is sensitive to the soul's sufferings.' Wherefore the physician heals and regulates the soul through the body; but the musician amends and controls the body through the soul. Accordingly, whoever was able to fulfill, in a variety of ways, the office of both physician and musician could govern the bodies and minds of men almost according to his wish. But this, surely, is to be treated as a secret by discreet philosophers.

XXV

The rays of all stars are double, some sensible or luminuous, others of more secret influence. The latter penetrate in an instant of time everything that is contained in the universe; the former can be prevented by some means from penetrating so far.

As Peter French rightly points out: 'Therefore, Dee concludes, the world is like a lyre. He explains that the overall structure of the universe, its harmonies and dissonances, sympathies and antipathies, determines the sweet and infinite variety of the marvellous music drawn from the individual strings. This is the cardinal proposition of Dee's *Propaedeumata Aphoristica*. It is the cardinal proposition of all Renaissance magical philosophy.'[5]

According to the *Aphorisms*, wonders can be worked by means of a union between celestial powers and the human imagination. Dee goes on to outline the nature of the Four Elements of ancient philosophy and the means of their manipulation, comparing invisible natural forces with the visible powers of the magnet, which has effects at a distance and affects matter with its rays. He insists on the benefits to be gained by more accurate astronomical observation, for a thorough and correct comprehension of the behaviour of celestial forces will enable Man to achieve more with his science than Nature alone and unaided.

We may leave the *Aphorisms* remembering Heilbron's summary: 'They echo the harmonies of the world and the sympathies among all things, and dimly announce the hermeticism that was to inform the *Monas Hieroglyphica*.'[6]

3

TO SIR WILLIAM CECIL
[1563]

WHAT was the harmony for which Dee sought? As Frances Yates has demonstrated,[1] this question cannot be answered without a study of Renaissance Neoplatonism, Hermeticism, and Christian Cabalism; for, as she argues: 'the dominant philosophy of the Elizabethan age was precisely the occult philosophy, with its magic, its melancholy, its aim of penetrating into profound spheres of knowledge and experience, scientific and spiritual, its fear of the dangers of such a quest, and of the fierce opposition which it encountered.'[2] And this 'occult philosophy' was 'compounded of Hermeticism as revived by Marsilio Ficino, to which Pica della Mirandola added a Christianised version of Jewish Cabala. These two trends, associated together form . . . "the occult philosophy", which was the title which Henry Cornelius Agrippa gave to his highly influential handbook on the subject.'[3]

This 'occult philosophy' was propagated by Pico and Ficino in late fifteenth-century Italy. It was an amalgam of pure Platonism; Neoplatonism as expressed in the work of men such as Plotinus and Iamblichus; Pythagorean numerology, Gnosticism, Chaldean lore ascribed to Zoroaster, and medieval magic deriving from Roger Bacon and Albertus Magnus; the 'Art' of the philosopher and mystic, Ramon Lull; and a body of 'Hermetic' writings on mysticism and magic attributed to 'Hermes Trismegistus', a legendary Egyptian magus believed to have lived at about the same time as Moses. However, Pico parted intellectual company from Ficino by introducing Cabala, which he lauded in his *Conclusions* as 'confirming the Christian religion from the foundations of Hebrew wisdom'.[4]

'What is Cabala?' asks Frances Yates, and answers: 'The word means "tradition". It was believed that when God gave the Law to Moses he gave also a second revelation as to the secret meaning of the Law. This esoteric tradition was said to have been passed down the ages orally by initiates. It was a mysticism and a cult but rooted in the text of the Scriptures, in the Hebrew language, the holy language in which God had spoken to man.'[5]

The essence of 'the occult philosophy' can be stated in the following propositions:

1.   All is a Unity, created and sustained by God through His Laws.

2.   These Laws are predicated upon Number.

3.   There is an art of combining Hebrew letters and equating them with Number so as to perceive profound truths concerning the nature of God and His dealings with Man.

4.   Man is of divine origin. Far from being created out of dust, as in the *Genesis* account, he is in essence a star daemon.

5.   As such, he has come from God and must return to Him.

6.   It is essential to regenerate the divine essence within Man, and this can be done by the powers of his divine intellect.

7.   According to the Cabala, God manifests by means of ten progressively more dense emanations: and Man, by dedicating his mind to the study of divine wisdom, by refining his whole being, and by eventual communion with the angels themselves, may at last enter into the presence of God.

8.   An accurate understanding of natural processes, visible and invisible, enables Man to manipulate these processes through the powers of his will, intellect and imagination.

9.   The Universe is an ordered pattern of correspondences.

'The belief in the manipulatory ability of man is all important. The revival of Hermeticism marks the dawn of the scientific age because it unleashed the driving spirit that inspired man to compel natural forces to serve him to an extent never dreamed of before,' writes Peter French, and adds: 'It was not the universe that had changed for the Renaissance magus; it was the role of man that was perceived anew.'[6]

The matter was taken further by Henry Cornelius Agrippa (1486–1535), whose life, work, contacts, and mysterious travels cry out for more scholarship. Agrippa urged that the universe be conceived as consisting of three worlds: the world of Elemental or Terrestrial Nature, which was the province of the physical sciences; the Celestial World of the stars, which could be understood and manipulated by the study and practice of astrology and alchemy; and the Supercelestial World, which could be entered into and apprehended by numerical operations and the conjuring of the angels themselves.

Agrippa was prepared to go further than his predecessors, as Frances Yates has stated: 'It is the Ficinian magic which Agrippa teaches in his first book, though he teaches it in a much bolder way. Ficino was nervous of the magic; he was anxious to keep his magic "natural", concerned only with elemental substances in their relations to the stars, and avoiding the "star demons", the spirits connected with the stars. It was really not possible to teach astral magic whilst avoiding the star demons, as Agrippa saw and boldly accepted the challenge.'[7]

Theory was not enough for Agrippa. One of his primary concerns was with Ceremonial Magic and its technical procedures. In other words, he taught how to *do* it.

Others who can be placed within this tradition of thought and/or practice include the great scholar of the German Renaissance Johannes Reuchlin; Francesco Giorgi, the Cabalist friar of Venice; Paracelsus, the magician, scientist, and founder of Western alternative medicine; the benedictine Abbot John Trithemius; and John Dee.

About December 1562, Dee went to Antwerp to arrange with the city printers certain matters concerning the publication of some of his books. Events took place which caused him to write a letter *To Sir William Cecil* on 16 February 1563, and this forms the subject matter of the next extract. Dee asks Cecil for permission to extend his stay. Why should this be required, and from no less a personage than Queen Elizabeth's chief minister? Some interesting explanations have been offered, which will be considered later.

Dee's request was based primarily upon his acquisition of *Steganography* by Abbot Trithemius forty years before its first printing. It is this work which Dee eulogizes in his letter to Cecil, though subsequent commentators have differed regarding its true nature. John E. Bailey summarized it simply as 'the first studied work on cipher writing':[8] but D. P. Walker wrote: 'I believe that trithemius' *Steganography* is partly a treatise on cryptography in which the methods of encipherment are disguised as demonic magic, and partly a treatise on demonic magic.'[9]

One wonders what Cecil made of it all. After all, he was always the practical politician. According to Isaac D'Israeli,[10] he feared that Dee might go mad. Yet in the fifth chapter of Dee's *Compendious Rehearsal*, a certificate is mentioned, dated 28 May 1563, 'in which that statesman testifies that Dee's time beyond seas had been well bestowed'.[11]

It has been argued that John Trithemius (1462–1516) exercised a profound influence upon the thought of Dee. Some have perceived vital links between *Steganography* and Dee's *Monas Hieroglyphica*, written in the following year and printed at Antwerp, and it is worth quoting the opinion of Dr Robert Hooke: 'Now tho' at that time the key or method of that book [*Steganography*] were not so well and commonly known, yet I do not doubt but this inquisitive man had got knowledge of it in his travels and enquiries in Germany, possibly when he presented his *Monas Hieroglyphica* to the Emperor Maximilian in 1564.'[12] This is pointed out by Bailey, who also refers the reader to note R in the *Biographia Britannica* life of Dee (ed. 1750, vol. iii, 1644–5) as treating of the connection.

There is a further mystery in the ensuing letter. Dee stresses that he has met '*such* men . . . as I'd never hoped to have in terms of assistance', and declares of Trithemius' book: 'The understanding of it I doubt not to attain, by God's Grace and by conference with such men as are already down

in my diary: men hard to find, although daily seen.'
  Who were they?

Right honorable Sir,

  My most humble obeisance in due sort considered, it may please
you to understand that the approved wisdom, with which the Almighty
has endowed you; and the exact balance of justice whereby mens' doings
in your hands are ordered; and the natural zeal as well to good letters
(which from your tender age has in your breast continually increased)
as to the honour and public weal of our Country (which now in you
freshly flowers and yields fruit abundantly): These, and other
considerations, have directed my choice to you only among so many
others in places of high honour and governance: choice, I say, whereby
your wisdome, justice and aforesaid zeal may (if so it stands with your
good will) be transplanted to far flung lands and strange peoples: if
my hand be not unlucky in guiding so weighty a charge.
  Therefore, briefly to place before your eyes the chief request on
which my case stands: Albeit that our universities both have in them
men right excellent in sundry branches of knowledge as in Divinity,
the Hebrew, Greek and Latin tongues, &c. Yet for as much as the
Infinite Wisdom of our Creator is branched into manifold more sorts
of wonderful Sciences, greatly aiding divine insights into a better view
of His Power and Goodness, our Country has no man (that I have
ever heard of) able to set his foot or show his hand in the *Science De
Numeris formalibus*, the *Science De Ponderibus mysticis* and the *Science
De Mensuris divinis* (by which three the huge frame of this world is
fashioned compactly, reared, established and preserved): and in other
Sciences, either collateral with them, or derived from them, or prompted
by them.
  And some such knowledge, after my long search and study, great
cost and travail, has (through God's Mercy and Grace) fallen under
my perseverence and understanding (whereof I am taught to render
account with increase of my talent) and so I have forced my wit and
pawned my self to draw together and disclose by writing, such profitable
and pleasant Sciences. And wasting no time (the frailty of life and
health being as it is), I thought it good, this season of Christmas festivals
(commonly otherwise spent) to make a start to Antwerp, and there
to employ that time in taking and setting orders with sundry Dutch
printers and other artificers, for the true and diligent printing of such

my labours as I have by me ready for the press; and thereupon I intended forthwith to return before Easter at the latest, because I hoped to have found things and men apt to my purpose.

But lo! it so falls out now that I cannot compass this my intent but am driven to deal with printers of High Germany, whereby a longer time will run. And also it seems my coming here (see, I pray you) is almost incredible, for by diligent searching and travail (for so short a time!) *such* men and such books have come to my knowledge concerning the aforementioned Great Sciences, as I'd never hoped to have in terms of assistance, either from one or the other.

So that in most reverent wise (the premises considered), I make to your Honour my humble petition: That you will charitably inform me of your pleasure and advise and counsel me whether you will have me return forthwith, my books unprinted and out of my hands; and also disdaining and neglecting this offer and occasion by the hand of God, whereby His glory, the honour of your good self and (so may it chance) the weal of my country may be advanced.

Or that you will herein declare your wisdom, justice and zeal (which in many cases far inferior to this, you have not withdrawn) in procuring leave, yea, and aid to my small ability to abide the better while achieving so great a feat as (by the enjoying of these men and books) by God's leave I intend to essay.

And for a proof more evident of my endeavour and purpose, it may please you to understand that already I have purchased one book, for which a thousand crowns have been by others offered, and yet it could not be obtained. A book for which many a learned man has long sought and daily yet does seek: whose use is greater than the fame spread about it; the name is not unknown to you. The title is *Steganographia* by John Trithemius, concerning which mention is made in both the editions of his *Polygraphia*, and in his epistles, and in sundry other men's books. A book for your Honour or a Prince, so fit, so needed and commodious in human knowledge that none can be fitter or more worthy.

Of this book, the one half have I copied out, with continual labour over the most part of ten days: and now I am subject to the patronage of a nobleman of Hungary for the writing of the rest: he has promised me leave for this after he perceives that I shall be remaining with him longer (with the leave of my prince) to pleasure him also with such points of Science as he may require at my hands.

I assure you, the means that I used to compass the knowledge of where this man and others like him are, and likewise of similar books as at present I am aware of, have cost me all that ever I could honestly borrow here, beside that which I thought needful to bring with me for so short a time, to the value of XX^{li}. God knows my zeal for honest and true knowledge, for which my own flesh, blood and bones would be the merchandise if the case so required. This book, either as I now have it, or hereafter shall have it, fully whole and perfect (if it please you to accept my present) I give to Your Honour as the most precious jewel that I have yet recovered from other men's travails. The understanding of it I doubt not to attain, by God's Grace and by conference with such men as are already down in my diary: men hard to find, although daily seen.

And then I shall also think Your Honour most worthy of it, for procuring me *dulcia illa ocia*; the fruit whereof my Country *et tota Resp. Literaria* justly shall ascribe to your wisdom and honourable zeal towards the advancement of good Letters and wonderful, divine and secret Sciences. And whatever your will and order with me shall be, I will request some of my friends to resort to Your Honour to understand, this case being being as strange to them as it is to me and contrary to expectations. As knows the Almighty who preserves Your Honour with continuance of health and abundance of His Grace according to his good pleasure.

[From *Dee and Trithemius's 'Steganography'*, ed. and introduced by John E. Bailey, *Notes And Queries*, fifth series, 24 May 1879. Transliteration: GS.]

4

MONAS HIEROGLYPHICA
[1564]

THE *Monas Hieroglyphica* (1564), dedicated to the Emperor Maximilian II, is Dee's most celebrated yet most inaccessible work. Dee wrote it in just thirteen days, 13–25 January 1564 in Antwerp, though he tells us it was the result of 'seven years' gestation'. For three centuries it could be read only in the original Latin: an English translation by J. W. Hamilton-Jones, with an accompanying commentary, was published in 1947 and republished in 1975 and 1977;[1] a translation and commentary by C. H. Josten appeared in 1964.[2]

What is one to make of the *Hieroglyphic Monad*? Even Frances Yates confessed that the explanatory text 'leaves the reader thoroughly bewildered'.[3] Commentators agree that the key is no longer with us, that key being Dee's oral explanation; or perhaps we are too far removed from sixteenth-century intellectual sensibilities to perceive implications deeply significant to intelligent men of that time. The *Hieroglyphic Monad* consists of the primary symbol itself and the acompanying text of twenty-four Theorems with diagrams. Certainly Dee regarded it as his masterpiece, the summary and crowning synthesis of all the knowledge and wisdom he had acquired. As Peter French writes: 'He thought that through anamnesis, he had perhaps been able to discover within himself the secrets of the ancient magi and develop them exactly as his spiritual ancestors would have wished.'[4] In Theorem XXIII Dee wrote: 'In the name of Jesus Christ crucified upon the Cross, I say the Spirit writes these things rapidly through me: I hope, and I believe, I am merely the quill which traces these characters.'

The *Hieroglyphic Monad* is Dee's endeavour to create a unifying symbol which embodies the entire cosmos. But it is not enough just to study the symbol and the text intellectually. The glyph must be meditated upon so deeply and so often that it becomes engraved within the psyche, part of the very fabric of one's being, thus bringing about the divine regenerative experience so eagerly sought after by the occult philosophers.

Dee was not alone in his ambition to summarize the All in the One. Other symbols make the same attempt: the Yin-Yang symbol, from which the I-Ching derives, for instance; or the Tree of Life glyph, which represents and incorporates the Cabala and of which Dee must have been fully aware. [5] He intended, however, to create a new, coherent, practical cabalistic synthesis which would revolutionize thought.

As Diane de Prima writes in her Introduction to the 1975 edition:

it is a diagram, at once, of process and goal. From the point in the centre of the circle, the entire glyph unfolds, theorem by theorem . . . it is expressive of mathematical relation as universally applicable as ''e = mc² '' and Dee sets it to work on many different levels of learning. In his dedicatory letter to Maximilian II, he states that his book will re-organise the science of the grammarians, reveal a new notation of number, revolutionise geometry and logic, make obsolete the present practice of music, optics and astronomy, and broaden for both the cabalist and the philosopher the understanding of his art. [6]

It is here too that we find a concern which came increasingly to preoccupy Dee: alchemy. This word has often been used to mean the attempts to turn base metals into gold or create the Elixir of Life or the Philosopher's Stone which led to the development of modern chemistry; but the aim of the true alchemist was the divine transmutation of the interior self, of which the transmutation of mineral substances was but the outward sign of an inner grace.

THEOREM I

It is by the straight line and the circle that the first and most simple example and representation of all things may be demonstrated, whether such things be either non-existent or merely hidden under Nature's veils.

THEOREM II

Neither the circle without the line, nor the line without the point,

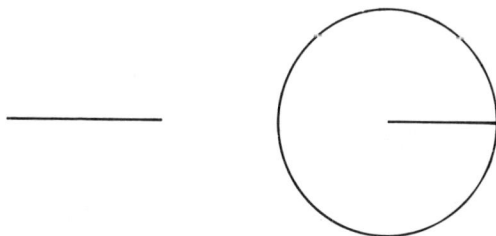

can be artificially produced. It is, therefore, by virtue of the point and the Monad that all things commence to emerge in principle.

That which is affected at the periphery, however large it may be, cannot in any way lack the support of the central point.

THEOREM III

Therefore, the central point which we see in the centre of the hieroglyphic Monad produces the Earth, round which the Sun, the Moon, and the other planets follow their respective paths. The Sun has the supreme dignity, and we represent him by a circle having a visible centre.

The Hieroglyphic Monad

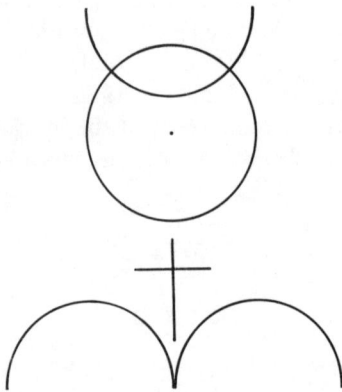

Note: at this point it is interesting to note the reaction of Queen Elizabeth to the *Monas* as described by Dee in *The Compendious Rehearsal*:

After my return from the Emperor's court, Her Majesty very graciously vouchsafed to account herself my scholar of my book, written to the Emperor Maximilian, entitled *Monas Hieroglyphica*; and said, whereas I had prefixed in the forefront of the book; *Qui non intellegit, aut taceat, aut discat* [whoever does not understand should either learn or be silent]: if I would disclose to her the secrets of that book, she would *et discere et facere*; whereupon Her Majesty had a little perusing of the same with me, and then in a most heroic and princely manner did comfort me and encourage me in my philosophical and mathematical studies . . .

THEOREM XIV

It is therefore clearly confirmed that the whole magistery depends upon the Sun and the Moon. Thrice Greatest Hermes has repeatedly told us this in affirming that the Sun is its father and the Moon is its mother: and we know truly that the red earth (*terra lemnia*) is nourished by the rays of the Moon and the Sun which exercise a singular influence upon it.

THEOREM XV

We suggest, therefore, that Philosophers should consider the action of the Sun and the Moon upon the Earth. They will notice that when the light of the Sun enters Aries, then the Moon, when she enters the next sign, that is to say Taurus, receives a new dignity in the light and is exalted in that sign in respect of her natural virtues. The Ancients explained this proximity of the luminaries — the most remarkable of all — by a certain mystic sign under the name of the Bull. It is very certain that it is this exaltation of the Moon to which in their treatises the astronomers from the most ancient times bear witness. This mystery can be understood only by those who have become the Absolute Pontiffs of the Mysteries. For the same reason they have said that Taurus is the house of Venus — that is to say, of conjugal love, chaste and prolific, for nature rejoices in nature, as the great Ostanes concealed in his most secret mysteries. These exaltations are acquired by the Sun, because he himself, after having undergone many eclipses of his light, received the force of Mars, and is said to be exalted in this same house of Mars which is our Ram (Aries).

This most secret mystery is clearly and perfectly shown in our

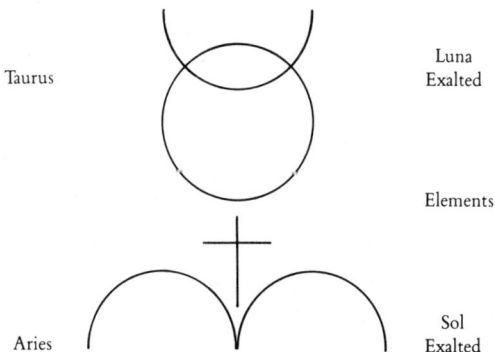

Taurus

Luna
Exalted

Elements

Aries

Sol
Exalted

Monad by the hieroglyphic figure of Taurus, which is here represented, and by that of Mars, which we have indicated in Theorem XII and Theorem XIII by the Sun joined to a straight line towards the sign of Aries.

Note: we are now grappling with the Monad as the key to astrological and alchemical mysteries. C. H. Josten's *Introduction* remains the best guide here.

Aries is the first sign of the Zodiac and is thought to correspond with the Element of Fire: while Taurus is the second and is thought to correspond with the Element Earth. The Cross symbolizes the Four Elements of the ancients, the other two being Water and Air. Above the symbol of Aries is that of the planet Mercury; and if the semi-circle at the top is removed, it is transformed into the symbol of the planet Venus; astronomically, Mercury is the planet closest to the Sun, then Venus, and then Earth. In the final rescension of the Monad, it is enclosed within an egg; this is the Philosopher's Egg or alchemical vessel within which the transmutation takes place.

More meditation will reveal more correspondences and set up interesting trains of thought. Diane de Prima has given us some of her own investigations in her *Introduction*. Dee himself was convinced that this exercise was worthwhile, for it would lead to extraordinary results. He wrote: 'He who fed [the monad] will first himself go away into a metamorphosis and will afterwards very rarely be held by mortal eye. This . . . is the true invisibility of the *magi* which has so often (and without sin) been spoken of, and which (as all future *magi* will own) has been granted to the theories of our monad.'[7]

Certain contemporaries esteemed the work highly, and it was published again in Frankfurt in 1591. Moreover, the second *Rosicrucian Manifesto*, printed 1614–15, includes a discourse on secret philosophy based on *The Hieroglyphic Monad*, and the sign appears in *The Chemical Wedding*.[8]

It is possible to read a sexual interpretation into Theorem XV, as in the case of many alchemical texts. Some might consider that this interpretation is confirmed by a passage in Dee's introductory letter to Maximilian:

'There is present, hidden in the most central point of our Hieroglyphic Monad, a terrestrial[9] body. How this body may be activated by Divine force, the monad teaches without words. When activated, it copulates (in a perpetual marriage) with the sun and the moon — even if before this, whether in heaven or elsewhere, the sun and moon were completely separate from this body. When the marriage has been performed . . . [the] truly great, metaphysical Revolution is completed.'

In the following and final extract, Dee writes of number and relation, wonders whether he has sinned in revealing such a great mystery, and appeals urgently to Maximilian and his descendants to understand the mystery so as to restore the honour of Christ upon Earth.

THEOREM XX

We have shown sufficiently that for very good reasons the Elements are represented in our Hieroglyph by the straight lines, therefore we give a very exact speculation concerning the point which we place in the centre of our Cross. This point cannot by any means be abstracted from our Ternary. Should anyone who is ignorant of this divine learning, say that in this position of our Binary the point can be absent, we reply, he may suppose it to be absent, but that which remains without it will certainly not be our Binary; for the Quaternary is immediately manifested, because by removing the point we discontinue the unity of the lines. Now, our adversary may suppose that by this argument we have reconstructed our Binary; that in fact our Binary and our Quaternary are one and the same thing, according to this consideration, which is manifestly impossible. The point must of necessity be present, because with the Binary it constitutes our Ternary, and there is nothing that can be substituted in its place. Meanwhile he cannot divide the hypostatic property of our Binary without nullifying an integral part of it. Thus it is demonstrated that it must not be divided. All the parts of a line are lines. This is a point, and this confirms our hypothesis. Therefore, the point does not form part of our Binary and yet it forms part of the integral form of the Binary. It follows that we must take notice of all that is hidden within this hypostatic form and understand that there is nothing superfluous in the linear dimension of our Binary. But because we see that these dimensions are common to both lines, they are considered to receive a certain secret image from this Binary. By this we demonstrate here that the Quaternary is concealed within the Ternary. O God! pardon me if I have sinned against Thy Majesty in revealing such a great mystery in my writings which all may read, but I believe that only those who are truly worthy will understand.

We therefore continue to expound the Quaternary of our Cross as we have indicated. Seek diligently to discover whether the point may be removed from the position in which we first find it. The mathematicians teach that it may be displaced quite simply. At the moment when it is separated the Quaternary remains, and it becomes much more clear and distinct to the eyes of all.

This is not a part of its substantial proportions, but only the confused and superfluous point which is rejected and removed.

O Omnipotent Divine Majesty, how we Mortals are constrained to confess what great Wisdom and what ineffable mysteries reside in the Law which Thou hast made! Through all these points and these letters the most sublime secrets, and terrestrial arcane mysteries, as well as the multiple revelations of this unique point, now placed in the Light and examined by me, can be faithfully demonstrated and explained. This point is not superfluous within the Divine Trinity, yet when considered, on the other hand, within the Kingdom of the four Elements it is black, therefore corruptible and watery. O thrice and four times happy, the man who attains this (almost copulative) point in the Ternary and rejects and removes that sombre and superfluous part of the Quaternary, the source of vague shadows. Thus after some effort we obtain the white vestments brilliant as the snow.

Oh, Maximilian! May God, through this mystagogy, make you or some other scion of the House of Austria the most powerful of all when the time comes for me to remain tranquil in Christ, in order that the honour of His redoubtable name may be restored within the abominable and intolerable shadows hovering above the Earth. And now for fear that I myself should say too much I shall immediately return to the burden of my task, and because I have already terminated my discourse for those whose gaze is centred within the heart, it is now necessary to translate my words for those whose heart is centred within their eyes.

The recommendations of Diane de Prima are of interest: 'The real key to this book seems to me to be inherent in the glyph itself: draw the Monad frequently, look long at it, use it in your meditations, and slowly it begins to speak. Read the text with an open mind and even more, an open heart [10] . . . it is by working thus with the glyph till we have exhausted our intuition and then referring again to the text, "reading between the lines", and then working again with the glyph, that, hopefully, the full meaning of this work will re-surface for us.' [11]

Possibly so. In any case, the reader will react to *The Hieroglyphic Monad* in one of several ways: he will dismiss it altogether; he will retire in dismay at its incomprehensibility; he will enjoy a curious item in the history of ideas; he will appreciate its point but object to the difficulties of study on the grounds that other ways to the same goal are easier and just as good: or he may choose to echo Dee's admonition on the Frontispiece.

'Who does not understand should either learn or be silent.'

5

THE PREFACE TO EUCLID
[1570]

ALTHOUGH Shumaker and Heilbron perceive in Dee 'a continuous progress toward the occult and irrational',[1] this is not evident in Dee's next important work, the *Mathematical Preface* to Euclid's *Elementes of Geometrie,* first translated into English by Sir Henry Billingsley (London, 1570). This *Preface* was so highly respected it was reprinted in the editions of 1651 and 1661, almost a century later.[2]

'The subject of the Preface is the importance of number and of the mathematical sciences, and this is confirmed by quotation from one of Pico della Mirandola's Mathematical Conclusions: "By number, a way is had, to the searching out and understanding of every thyng, hable to be known",' writes Frances Yates.[3] Dee endeavours to show how all sciences ultimately derive from number and is once again creating a synthesis whereby the All can be expressed in the One. In this case, Dee uses a diagram which he terms the *Groundplat,* in which he not only tried to demonstrate the coherence and essential unity of mathematical and scientific knowledge, but also defined the 'Sciences and Artes Mathematicall' so as to summarize the state of learning in his time.

The second extract, *The Arte of Navigation,* demonstrates knowledge of a subject which would bear even fuller fruit in *General And Rare Memorials Pertaining To the Perfect Art of Navigation* (1577). According to E. G. R. Taylor in *Tudor Geography:*[4] 'Dee gives the first English definition of the art of navigation (excluding Richard Eden's translation of Martin Cortes's *Art of Navigation,* 1561) and one that is hard to surpass.' I have left it in the original.

The final extract is a splendid piece of invective against slanderers and an impassioned refutation of rumour. Some might argue that here Dee displays a disconcerting paranoia, for he certainly believed that he had enemies who consistently defamed his character and the nature of his work. Others may argue that the passage is as true of genuine seekers after wisdom in the twentieth century as it was for Dee four hundred years ago.

Perhaps the fairest verdict on the Preface as a whole is still that of Taylor in *Tudor Geography:* 'a magnificent exposition of the relationship and application of mathematics, especially arithmetic and geometry, to the practice of various skilled arts and crafts.'

(i)

THE GROUND-PLAN OF THE MATHEMATICAL PREFACE
OF MR JOHN DEE

Sciences and Arts Mathematical are either:

1   Principal, of which are two only:

A —   Arithmetic, Simple or Mixed.

B —   Geometry, Simple or Mixed.

1Ai — *Arithmetic Simple*
Which deals with Numbers only and demonstrates all their properties and appurtenances, where a Unit is indivisible.

1Aii — *Arithmetic Mixed*
Which with aid of Geometry Principal, demonstrates some Arithmetical conclusion or purpose.

1Bi — *Geometry Simple*
Which deals with Magnitudes only, and demonstrates all their properties, passions and appurtenances.

1Bii — *Geometry Mixed*
Which with aid of Arithmetic Principal, demonstrates some Geometrical purpose, such as Euclid's *Elements.*

THE USE WHEREOF IS EITHER:

1I —   In things supernatural, eternal and divine, by application ascending.

1II —   In things mathematical, without further application.

1III — In things natural, both substantial and accidental, visible and invisible & c., by application descending.

The like uses and applications (though in a degree lower) in the arts Mathematical derivative.

OR Sciences and Arts Mathematical are:

2   Derivative from the Principles of which some have:

A —   The names of the Principals, or

B —   Proper names.

The names of the Principals are:

2Ai — *Arithmetic Vulgar,* which considers

I — The Arithmetic of most usual Numbers and of fractions to them appertaining.
II — The Arithmetic of Proportions.
III — Arithmetic Circular.
IV — The Arithmetic of Radical Numbers, Simple, Compound, Mixed and their Fractions.
V — The Arithmetic of Cossick Numbers with their Fractions. And the great Art of Algebra.

2Aii — *Geometry Vulgar*, which teaches measuring: by distance of the thing measured as:
I — How far from the Measurer any thing is of him seen on Land or Water, called APOMECOMETRY.
II — How high or deep from the level of the Measures standing, anything seen of him on Land or Water, called HYPSOMETRY.
II — How broad a thing is, which is in the Measurer's view to be situated on Land or Water, called PLATOMETRY.

Of which are grown the feats and arts of:
a — GEODESIE, more cunningly to measure and survey lands, woods, waters &c.
b — GEOGRAPHY
c — CHOROGRAPHY
d — HYDROGRAPHY
e — STRATARITHEMETRY
AND

2Aii — *Geometry Vulgar*, which teaches measuring at Hand
I — All lengths — MECOMETRY
II — All Planes, as Land, Board, Glass &c. — EMBADOMETRY.
III — All Solids, as Timber, Stone, Vessels &c. — STEREOMETRY.

2B Some Sciences and Arts Mathematical have Proper Names, as:
i — *Perspective*, which demonstrates the manners and properties of all Radiations, Direct, Broken and Reflected.
ii — *Astronomy*, which demonstrates the Distances, Magnitudes and all Natural motions, appearances and passions proper to the planets and fixed stars for any time past, present and to come: in respect of a certain Horizon, or without respect thereof.
iii — *Music*, which demonstrates by reason and teaches by sense

perfectly to judge and order the diversities of Sound, high and low.

iv — *Cosmography,* which wholly and perfectly makes description of the Heavenly and also Elemental part of the World: and of these parts makes homological application and mutual collation necessary.

v — *Astrology,* which reasonably demonstrates the operations and effects of the natural beams of light, and secret influence of the planets and fixed stars in every Element and Elemental Body, at all times in any Horizon assigned.

vi — *Statike,* which demonstrates the causes of heaviness and lightness of all things, and of the motions and properties to heaviness and lightness belonging.

vii — *Anthrography,* which demonstrates the number, measure, weight, figure, situation and colour of every divers thing contained in the perfect body of man, and gives certain knowledge of the figure, symmetry, weight, characterisation and due local motion of any parcel of the said body assigned, and of numbers to the said parcel appertaining.

viii — *Trochelike,* which demonstrates the properties of all circular motions: simple and compound.

ix — *Helicosophy,* which demonstrates the designing of all spiral lines, in plain, on Cylinder, Cone, Sphere, Conoid and Spheroid, and their properties.

x — *Pneumatithme,* which demonstrates by close, hollow geometrical figures (regular and irregular) the strange properties (in motion or stay) of the Water, Air, Smoke and Fire, in their Continuity and as they are joined to the Elements next to them.

xi — *Menandy,* which demonstrates how above Nature's virtue and power simple, virtue and force may be mutliplied in order to direct, to lift or to pull to; and to put or cast from any multiplied or simple determined virtue, weight or force naturally not so directable or moveable.

xii — *Hypogeidy,* which demonstrates how under the spherical superfices of the Earth, at any depth to any perpendicular line assigned (whose distance from the perpendicular of the entrance and the Azimuth likewise, in respect of the said entrance, is known with certainty) may be prescribed.

xiii — *Hydragogy,* which demonstrates the possible leading of Water by Nature's Law, and by artificial help from any head (being spring, standing or running Water) to any other place assigned.

xiv — *Horometry,* which demonstrates how at all times appointed, the precise denomination of time may be known, for any place assigned.

xv — *Zography,* which demonstrates and teaches how the intersection of all visual Pyramids made by any plane assigned (the centre, distance and lights being determined) may be by lines and proper colours represented.

xvi — *Architecture,* which is a Science garnished with many doctrines and divers instructions: by whose judgement all works finished by others, are judged.

xvii — *Navigation,* which demonstrates how by the shortest route, by the aptest direction and in the shortest time, a sufficient ship between any two places (in a navigable passage) assigned, may be conducted: and in all forms of change and natural disturbances, how to use the best possible means to recover the place first assigned.

xviii — *Thaumaturgike,* which gives certain order to make strange works of the sense to be perceived and of men greatly to be wondered at.

xix — *Archimastry,* which teaches to bring to actual experience sensible, all worthy conclusions, by all the Arts mathematical purposed: and by true natural Philosophy concluded: and both adds to them a farther scope in terms of the same arts: and also by this proper method and in peculiar terms proceeds, with help of the aforesaid arts, to the performances of complete Experiences, which of no particular Art, are able (formally) to be challenged.

[Note. 'cussick' means that which pertains to Algebra, the 'azimuth' of a sun or star is an arch between the meridian of the place and any given vertical line. Transliteration: G.S.]

(ii)

THE ART OF NAVIGATION, demonstrateth how, by the shortest good way, by the aptest Directio, & in the shortest time, a sufficient Ship, betwene any two places (in passage Navigable,) assigned: may be coducted: and in all stormes, & natural disturbances chauncyng, how, to vse the best possible meanes, whereby to recouer the place first assigned. What nede, the *Master Pilote,* hath of other Artes, here before recited, it is easie to know: as of *Hydrographie, Astronomie, Astrologie,* and *Horometrie.* Pre-supposing continually, the common Base, and foudacion of all: namely. *Arithmeticke* and *Geometrie.* So that, he be hable to vnderstand, and Iudge his own necessary Instrumentes, and furniture Necessary: Whether they be perfectly made or no: and also can, (if nede be) make them, hym selfe. As Quadrants, the Astronomers Ryng, The Astronomers staffe, the Astrolabe vniversall. An Hydrographicall Globe. Charts Hydrographicall, true, (not with parallel Meridians). The Common Sea Compas: The Compas of variacion: The Proportionall, the Paradoxall Compasses (of me Invented, for our two Moscouy Master Pilotes, at the request of the Company) Clockes with spryngs: houre, halfe houre, and three houre Sandglasses: & sundry other Instrumetes: And also, be hable, on Globe, or Playne to describe the Paradoxall Compasse: and duely to vse the same, to all maner of purposes, whereto it was inuented, And also, be hable to Calculate the Planetes places for all tymes.

Moreouer, with Sonne Mone or Sterre (or without) be hable to define the Longitude & Latitude of the place, which he is in: So that, the Longitude & Latitude of the place, from which he sayled, be giuen: or by him, be knoune. whereto, appertayneth expert meanes, to be certified euer, of the ships way. &c. And by foreseing the Rising, Settyng, Nonestedying, or midnightyng of certaine tempestuous fixed Starres: or their Coniunctions, and Anglynges with the Planetes, &c. he ought to have expert coniecture of Stormes, Tempestes, and Spoutes: and such lyke meteorologicall effectes, daungerous on Sea. For (as Plate sayth,) *Mutationes, opportunitatisq temporum presentire. noli minus rei militari, quant Agriculturae, Nauigationiq conuenit. To foresee the alterations and opportunities of tymes, is convenient, no lesse to the Art of Warre, then to Husbandry, and Nauigation.* And besides such cunnyng meanes, more euident tokens in Sonne and Mone, ought of him to be knowen: such as (the Philosophical Poëte) *Virgilius* teacheth in hys *Georgikes.* Where he sayeth, . . . [Quotation from the 1st book of the Georgics, beginning 'Sol quoqq & exoriens . . .']

And so of Mone, Sterres, Water, Ayre, Fire, Wood, Stones, Birdes, and Beastes, and of many thynges els, a certaine Sympathicall forewarnyng may be had: some tymes to great pleasure and profit, both on Sea and Land. Sufficiently, for my present purpose, it doth appeare, by the premisses, how *Mathematicall,* the *Art of Nauigation,* is: and how it needeth and also vseth other *Mathematicall Artes:* And now, if I would go about to speak of the manifold Commodities, commyng to this Land, and others, by Shypps and *Nauigation,* you might thinke, that I catch at occasions, to vse many wordes, where no nede is.

Yet, this one thyng may I, (iustly) say. In Navigation, none ought to haue greater care, to be skillful, then our English Pylotes. And perchaunce, Some, would more attempt: And other Some more willingly would be aydyng, if they wist certainely, What Priuiledge, God had endued this Iland with, by reason of Situation, most commodious for *Nauigation,* to Places most Famous & Riche. And though (of Late) a young Gentleman, a Courragious Captaine, was in great readynes, with good hope, and great causes of persuasion, to have ventured, for a Discouerye, (eyther *Westerly,* by *Cape de Paramantia:* or *Esterly,* aboue *Notta Zemla,* and the *Cyremisses*) and was, at the very nere tyme of Attemptyng, called and employed otherwise (both then, and since,) in great good seruice to his Countrey, as the Irish Rebels haue tasted: Yet, I say, (though the same Gentleman, doo not hereafter, deale therewith) Some one, or other, should listen to the Matter: and by good aduise, and discrete Circumspection, by little, and little, wynne to the sufficient knowledge of that Trade and Voyage: Which, now, I would be sory, (through Carelesnesse, want of Skill, and Courage,) should remayne Vnknowne and vnheard of. Seyng, also, we are herein, halfe challenged, by the learned, by halfe request, published. Thereof, verely, might grow Commoditye, to this Land chiefly, and to the rest of the Christen Common wealth, farre passing all riches and worldly Threasure.

(iii)

A Refutation of Slander

And for these and such like marvellous Acts and Feats, naturally, mathematically and mechanically wrought and contrived, ought any honest student and modest Christian Philosopher be counted and called a CONJURER? Shall the folly of idiots and the malice of the scornful so much prevail that he who seeks no worldly gain or glory at their hands but only of God the treasure of heavenly wisdom and knowledge of pure hands: shall he (I say) in the mean space be robbed and spoiled of his honest name and fame.

He that seeks (by St Paul's advertisement) in the Creatures, Properties and wonderful Virtues to find just cause to glorify the Eternal and Almighty Creator: shall that man be (in hugger mugger) condemned as a companion of hell-hounds and a caller and Conjurer of wicked and damned spirits? He that bewails his great want of time sufficiently (to his contention) for learning of Godly wisdom and Godly verities and only therein sets all his delight: will that man waste and abuse his time in dealing with the chief Enemy of Christ our Redeemer; the deadly foe of all mankind; the subtle and impudent perverter of Godly Verity; the hypocritical crocodile, the envious basilisk, continually desirous in the twinkling of an eye to destroy all mankind both in body and soul eternally?

Surely (for my part, somewhat to say herein) I have not learned to make so brutish and so wicked a bargain. Should I for my twenty or twenty-five years study for two or three thousand marks spending; seven or eight thousand miles going; and travelling only for good learning's sake, and that in all manner of weathers, in all manner of ways and passages; both early and late; in danger of violence by man; in danger of destruction by wild beasts; in hunger and thirst; in perilous heats by day, with toil on foot; in dangerous damps of cold by night, almost bereaving life (as God knows; with lodgings, oft times, of small ease and sometimes of less security; and for much more (than all this) done and suffered for Learning and attaining of Wisdom: should one (I pray you) for all this, not otherwise, nor mere warily or (by God's Mercy) no more luckily have fished, with so large and costly a net, for so long a time in drawing, and that with the help and advice of Lady Philosophy and Queen Theology, but at length to have caught and drawn up a Frog? Nay, a Devil?

For so does the common peevish prattler imagine and jangle. And so does the malicious scorner secretly and bravely and boldly face down behind my back. Ah, what a miserable thing is this kind of man. How great is the blindness and boldness of the multitude in things above their capacity!

What a Land! What a People! What manners! What times are these? Are they become Devils themselves? And by false witness bearing against their neighbour, would they also become Murderers? Does God so long give them respite to reclaim themselves in from this horrible slandering of the guiltless, contrary to their own consciences, and yet will they not cease? Doth the innocent forbear the calling of them juridically to answer him according to the rigour of the Law? And will they despise his charitable patience?

As they, against him, by name do forge, fable, rage and raise slander by word and print: will they provoke him by word and print likewise to note their names to the world? With their particular devices, fables, beastly imaginings and unchristianlike slanders?

Well, well, (you such) my unkind Countrymen! O unnatural Countrymen! O unthankful Countrymen! O brainsick, rash, spiteful, and disdainful Countrymen! Why do you oppress me thus violently with your slandering of me, contrary to verity and contrary to your own consciences? And I, to this hour, neither by word, deed or thought have been in any way hurtful, damaging or injurious to you or yours? Have I so long, so dearly, so far, so carefully, so painfully, so dangerously fought and travelled for the learning of Wisdom and attaining of virtue, and in the end (in your judgement), am I become worse than when I began? A dangerous member of the Commonwealth — and no member of the Church of Christ? Call you this to be learned? Call you this to be a Philosopher and a lover of Wisdom? To forsake the straight, heavenly way and to wallow in the broad way of damnation. To forsake the light of heavenly wisdom and to lurk in the dungeon of the Prince of Darkness? To forsake the verity of God and His Creatures and to fawn upon the impudent, crafty, obstinate Liar and continual disgracer of God's Verity to the uttermost of his power? To forsake the Life and Eternal Bliss and to cleave unto the Author of Death everlasting? that murderous Tyrant, most greedily awaiting the Prey of Man's Soul.

Well: I thank God and Our Lord Jesus Christ for the comfort which I have by the examples of other men before my time; to whom neither in godliness of life nor in perfection of learning am I worthy to be compared; and yet they sustained the verie like injuries that I do: or rather greater. Patient SOCRATES, his *Apology* will testify to it: Apuleius, his *Apologies* will declare the brutishness of the multitude: Joannes Picus, Earl of Mirandola, his *Apologie* will teach you of the raging slander of the malicious ignorant against him: joannes Trithemius, his *Apologie* will specify how he had occasion to make public Protestations, as well by reason of the rude, simple folk, as also in respect of such as were counted to be of the wisest sort of men.

Many could I recite. But I defer the precise and determined handling of this matter, being loth to detect the folly and malice of my Native

Countrymen, who can hardly digest or like any extraordinary course of Philosophical studies not falling within the compass of their capacity, or where they are not made privy to the true and secret cause of such wonderful Philosophical Feats. These men are of four sorts chiefly:

The first I may name — VAIN PRATING BUSYBODIES. The second, FOND FRIENDS. The third, IMPERFECTLY ZEALOUS, and the fourth, MALICIOUS IGNORANT.

To each of these (briefly and in charity) I will say a word or two, and so return to my Preface.

VAIN PRATING BUSYBODIES: use your idle assemblies and conference otherwise than in talk of matter either above your capacities for hardness, or contrary to your consciences in truth.

FOND FRIENDS, leave off to commend your unacquainted friend upon blind affection. As, because he knows more than the common student, he must needs be skilful and a doer in such manner and matter as you term CONJURING. Weening thereby, you advance his fame; and you make for other men great marvels of yourselves to have such a learned friend. Cease to ascribe Impiety where you pretend Amity. For, if your tongues were true, then were your friend *Untrue,* both to God and his Sovereign. Such FRIENDS and FONDLINGS I shake off and renounce. Shake you off your Folly.

IMPERFECTLY ZEALOUS, to you do I say that (perhaps) you do mean well. But you miss the mark if you will kill a lamb to feed the flock with his blood. Sheep with lambs' blood have no natural sustenance. No more is Christ's rock duly edified with horrible slanders. Nor is your fair presence, by such rash, ragged Rhetoric, any whit well graced. But such as seek to use me, will find a foul crack in their credit. Speak of what you know. And know as you ought. Know not but hearsay when life lies in danger. Search to the quick and let charity be your guide.

MALICIOUS IGNORANT, what shall I say to thee? . . . CAUSE THY TONGUE TO REFRAIN FROM EVIL. RESTRAIN YOUR TONGUE FROM SLANDER.

[Transliteration: GS.]

6

GENERAL AND RARE MEMORIALS PERTAINING TO THE PERFECT ART OF NAVIGATION

[1577]

DEE'S greatest geographical contemporaries were Gerard Mercator, the greatest cosmographer, globe-maker and producer of navigational instruments in Europe and a close friend of Dee; Gemma Frisius, sometime cosmographer to the Emperor Charles V and with whom Dee was in contact; Abraham Ortelius of Antwerp, the renowned cosmographer, with whom Dee exchanged visits; Orontius Finaeus, Professor of Mathematics at the College de France, with whom Dee discussed science while in Paris in 1550 and 1551; and Pedro Nunez, Cosmographer Royal of Portugal and Professor of Mathematics at Coimbra, who in 1558 was appointed literary executor by Dee in the event of the latter's untimely death. The resulting exchanges of information proved to be milestones in the history of English geography, exploration and navigation.

Dee was the principal adviser in attempts to find a North-East passage to Cathay and a prime mover in the foundation of the Muscovy Company. Richard Chancellor was chosen to pilot the first voyage and Dee tutored him in navigation. Although the search for this North-East passage was a failure it opened up trade and communication between England and Muscovy, whose ruler, Ivan the Terrible, also became known by his courtiers as 'the English Tsar' on account of his Anglophilia. In fact, the Tsar became so impressed by Dee's reputation that he invited him to Muscovy, offering as inducement a large house, food prepared by his own chef, and the (then) splendid sum of £2000 a year. Dee declined for reasons of patriotism.

Martin Frobisher's attempts to find a North-West passage to the Orient in 1576, 1577 and 1578 were under Dee's guidance. Dee also advised Sir Humphrey Gilbert who, in September 1580, agreed to grant Dee rights to all lands his forthcoming voyage might discover north of the 50th parallel — in other words, the largest part of Canada. Unfortunately, this voyage was ill-fated and Gilbert was drowned.

1580 was also the year in which Dee presented a map of part of the Northern hemisphere to Queen Elizabeth I, endorsed with arguments setting out the legitimacy of England's rights in North America. Three years later Dee, with his last seafaring pupils Adrian Gilbert and John Davis, drew up a plan to carry out the colonization, conversion, and general exploitation of 'Atlantis', Dee's term for America. It is probable that Dee inspired Sir Walter Raleigh's expedition to the land he named 'Virginia' and the short-lived colony of Roanoke, 1584-90.

It is possible too that Dee was behind Sir Francis Drake's voyages around the globe: the evidence for this has been assembled in *Tudor Geography*[1] by Taylor, who writes: 'a close examination of the evidence leaves no doubt of (Dee's) intellectual honesty and genuine patriotism.'[2] Moreover: 'Recent researches make it clear that the credit for grasping the possibilities of arithmetical navigation, for doing the pioneer work on it, and for teaching its potentialities to both navigators and to younger mathematicians lies with Dr John Dee.'[3]

Nor can one ignore the technical innovations which won Dee the respect of English mechanicians. He devised the circumpolar chart in the 1550s, calculated a departure table, and designed a compass fly divided off into degrees as well as points.[4]

In the first extract from the *General and Rare Memorials pertayning to the Perfect Arte of Navigation* (London, 1577), Dee urges the creation of 'A Petty Navy Royal' in addition to a Grand Navy and states his case persuasively and cogently. In the second, he displays a practical cast of mind in his consideration of Fisheries.

The third extract is notable for a number of reasons, not least because it contains the first printed mention of 'The British Empire'. As Francis Yates points out: 'Expansion of the navy and Elizabethan expansion at sea were connected in his mind with vast ideas concerning the lands to which (in his view) Elizabeth might lay claim through her mythical descent from King Arthur.'[5]

The evidence compels one to agree with Taylor: '[Dee's] English Euclid and his efforts for the Reformation of the Calendar have won him an honoured place in the History of Mathematics. So, too, his unceasing efforts for the instruction of mariners, and for the unveiling of the hidden corners of the earth, entitle John Dee to an honoured place in the History of Geography.[6]

THE PETTY NAVY ROYAL

WHOM also I have heard often and most heartily wish, That all manner of persons passing or frequenting our seas appropriate, and many ways next environing England, Ireland, and Scotland, might be in convenient and honourable sort, at all times at the commandment and order, by beck or check, of a Petty Naval Royal of three-score tall ships or more, but in no case fewer; and they to be very well appointed, thoroughly manned, and sufficiently victualled.

The public commodities whereof ensuing are, or would be so great and many, as the whole commons, and all the subjects of this noble Kingdom would for ever bless the day and hour wherein such good and politic order was, in so good time and opportunity, taken and established: and esteem them not only most worthy and royal Councillors, but also heroical Magistrates, who have had so fatherly care for the commonalty; and most wisely procured so general British security,

1. That, henceforth, neither France, Denmark, Scotland, Spain, nor any other country can have such liberty for invasion, or their mutual conspiracies or aids, any way transporting, to annoy the blessed state of our tranquillity; as either they have in times past had, or else may have, whensoever they will forget or contemn the observing of their sworn or pretended amity.

2. Besides that, I report me to all English merchants, *said he,* of how great value to them, and consequently to the public weal of this Kingdom, such a security were? (a) Whereby, both outward and homeward, continually their merchantlike ships, many or few, great or small, may in our seas and somewhat further, pass quietly unpilled, unspoiled, and untaken by pirates or others in times of peace. (b) What abundance of money now lost by assurance [*marine insurance*] given or taken, would by this means also, be greatly out of danger?

3. And thirdly, (a) how many men, before time of urgent need, would thus be made very skilful in all the foresaid seas and sea coasts; in their channels knowing, in soundings all over, in good marks taking for avoiding dangers, in good harbours trying out, in good landings essaying, in the order of ebbs and floods observing, and all other points advisedly learning, which to the perfect Art of Navigation are very necessary: whereby they may be the better able to be divided and distributed in a greater Navy, with charge of Mastership or Pilotage, in time of great need. (b) They of this Navy should oftentimes espy or meet the privy sounders and searchers of our channels, flats, banks, pits, &c.; and so very diligently deciphering our sea coasts, yea, in the river of Thames also; otherwhile up to the station of the Grand Navy Royal. (c) And likewise, very often meet with the abominable thieves that steal our corn and victuals from sundry our coasts, to the great hindrance of the public plenty of England. And these thieves are both subjects and foreigners; and very often and to to [*far to*] evidently seen, and generally murmured at, but as yet not redressed; for all the

good and wise order by the most honourable Senate of the Privy Council taken therein.

4. Fourthly, how many thousands of soldiers of all degrees, and apt ages of men, would be, by this means, not only hardened well to brook all rage and disturbance of sea, and endure healthfully all hardness of lodging and diet there; but also would be well practised and easily trained up to great perfection of understanding all manner of fight and service at sea? so that, in time of great need, that expert and hardy crew of some thousands of sea soldiers [*Marines*] would be to this realm a treasure incomparable. And who knoweth not, what danger it is, in time of great need, either to use all fresh water soldiers; or to be a fortnight in providing a little company of *omni-gatharums,* taken up on the sudden to serve at sea? For our ordinary Land Musters are generally intended, or now may be spared to be employed otherwise, if need be.

5. How many hundreds of lusty and handsome men would be, this way, well occupied, and have needful maintenance, which now are either idle, or want sustenance, or both; in too many places of this renowned Monarchy?

6. Moreover, what a comfort and safeguard will it, or may it be to the whole Realm, to have the great advantage of so many warlike ships, so well manned and appointed for all assays, at all hours, ready to affront straightway, set on and overthrow, any sudden or privy foreign treachery by sea, directly or indirectly, attempted against this Empire, in any coast or part thereof. For *sudden* foreign attempts (that is to say, unknown or unheard of to us, before their readiness) cannot be done with great power. For great navies most commonly are espied or heard somewhat of, and that very certainly, while they are in preparing; though in the meanwhile, politicly, in divers places, they distribute their ships and their preparations appertaining.

7. And by reason of the foresaid Petty Navy Royal, it shall at all times, not only lie in our hands greatly to displease and pinch the petty foreign offender at sea; but also, if just occasion be given, on land to do very valiant service, and that speedily: as well against any of the foresaid foreign possible offenders, as also against such of Ireland or England, who shall or will traitorously, rebelliously, or seditiously assemble in troops or bands within the territories of Ireland or England; while greater armies, on our behalf shall be in preparing against them, if further need be. For skilful sea soldiers are also on land far more

trainable to all martial exploits executing; and therein to be more quick-eyed and nimble at handstrokes or scaling; better to endure all hardness of lodging or diet; and less to fear all danger near or far: than the land soldier can be brought to the perfection of a sea soldier.

8. By this Navy also, all pirates — our own countrymen, and they be no small number — would be called, or constrained to come home. And then (upon good assurance taken of the reformable and men of choice, for their good abearing from henceforth) all such to be bestowed here and there in the foresaid Navy. For good account is to be made of their bodies, already hardened to the seas; and chiefly of their courage and skill for good service to be done at the sea.

9. Ninthly, Princes and potentates, our foreign friends or privy foes, the one for love and the other for fear, would not suffer any merchant or others, subjects of the Queen's Majesty, either to have speedy wrong in their Courts; or by unreasonable delays or trifling shifts to be made weary and unable to follow their rights. And notwithstanding such our friends or privy foes, their subjects would be glad most reverently to become suitors and petitioners to the royal State of this Kingdom for just redress, if, any kind of way, they could truly prove themselves by any subject of this realm injured; and they would never be so stout, rude, and dishonourably injurious to the Crown and Dignity of this most sacred Monarchy as, in such cases, to be their own judges, or to use against this Kingdom and the royal chief Council thereof, such abominable terms of dishonour as our to to great lenity and their to to barbarous impudency might in a manner induce them to do. And all this would come to pass through the Royalty and Sovereignty of the seas adjacent or environing this Monarchy of England, Ireland, and (by right) Scotland and the Orkneys also, very princely, prudently, and valiantly recovered (that is to say, by the said Petty Navy Royal); duly and justly limited; discreetly possessed; and triumphantly enjoyed.

Should not Foreign Fishermen to injuriously abusing our rich fishings about England, Wales and Ireland) by the presence, oversight, power, and industry of this Petty Navy Royal be made content; and judge themselves well apaid to enjoy, by our leave, some great portion of revenue to enrich themselves and their countries by, with fishing within the seas appertaining to our ancient bounds and limits? Where now, to our great shame and reproach, some of them do come in a manner home to our doors; and among them all, deprive us yearly of many hundred thousand pounds, which by our fishermen using the said fishings as chief, we might enjoy; and at length, by little and little, bring them (if we would deal so rigorously with them) to have as little portion of our peculiar commodity (to our Islandish Monarchy, by GOD and Nature assigned) as now they force our fishermen to be contented with: and yearly notwithstanding, do at their fishing openly and ragingly use such words of reproach to our Prince and realm, as no true subject's heart can quietly digest. And besides that, offer such shameful wrongs to the good laboursome people of this land, as is not by any reason to be borne withal, or endured any longer: destroying their nets; cutting their cables to the loss of their anchors, yea, and oftentimes of barks, men and all.

And this sort of people they be, which otherwhile by colour and pretence of coming about their feat of fishing, do subtilly and secretly use soundings and searchings of our channels, deeps, shoals, banks, or bars along the sea coasts, and in our haven mouths also, and up in our creeks, sometimes in our bays, and sometimes in our roads, &c.; taking good marks, for avoiding of the dangers, and also trying good landings. And so, making perfect charts of all our coasts round about England and Ireland, are become almost perfecter in them, than the most part of our Masters, Leadsmen, or Pilots are. To the double danger of mischief in times of war; and also to no little hazard of the State Royal, if, maliciously bent, they should purpose to land any puissant army, in time to come.

And as concerning those fishings of England, Wales and Ireland, of their places, yearly seasons, the many hundreds of foreign fisherboats yearly resorting, the divers sorts of fish there taken, with the appurtenances: I know right well that long ago all such matter concerning these fishings was declared unto some of the higher powers of this Kingdom, and made manifest by R[OBERT]. H[ITCHCOCK]. another honest gentleman of the Middle Temple, who very discreetly and faithfully hath dealt therein; and still travaileth, and by divers other ways also, to further the weal public of England so much as in him lieth.

But note, I pray you, this point very advisedly. That as by this *Plat [plan]* of our said fishing commodities, many a hundred thousand pounds of yearly revenue might grow to the Crown of England more than now doth, and much more to the Commons of this Monarchy also: besides the inestimable benefit of plentiful victualling and relieving of both England and Ireland; the increasing of many thousands of expert, hard, and hardy mariners; the abating of the sea forces of our foreign neighbours and unconstant friends; and contrariwise, the increasing of our own power and force at sea; so it is most evident and certain that *principium* in this case is, *Plus quam dimidium totius,* as I have heard it verified proverbially in many other affairs.

Wherefore the very entrance and beginning towards our Sea Right recovering, and the foresaid commodities enjoying at length; yea, and the *only* means of our countinuance therewith, can be no other; but by the dreadrul presence and power, with discreet oversight and due order, of the said Petty Navy Royal; being — wholly sometimes, sometimes a part thereof — at all the chief places of our fishings; as if they were Public Officers, Commissioners, and Justiciers, by the supreme authority royal of our most renowned Queen ELIZABETH, rightfully and prudently thereto assigned.

So that this Petty Navy Royal is thought to be the only Master Key wherewith to open all locks that keep out or hinder this incomparable British Empire from enjoying, by many means, such a yearly Revenue of Treasure, both to the Supreme Head and the subjects thereof — as no plat [*tract*] of ground or sea in the whole world else, being of no greater quantity — can with more right, greater honour, with so great ease and so little charges, so near at hand, in so short time, and in so little danger, any kind of way, yield the like to either King or other potentate and absolute Governor thereof whosoever. Besides, the Peaceable Enjoyment, to enjoy all the same, for ever; yea, yearly and yearly, by our wisdom and valiantness duly used, all manner of our commodities to arise greater and greater; as well in wealth and strength as of foreign love and fear, where it is most requisite to be: and also of Triumphant Fame the whole world over, undoubtedly.

Also, this Petty Navy Royal will be the perfect means of very many other and exceeding great commodities redounding to this Monarchy; which our fishermen and their fisher-boats only, can never be able to compass or bring to pass: and those being such as are more necessary to be cared for presently [*instantly*] than wealth.

Therefore, the premises well weighed, above and before all other, this Plat of a Petty Navy Royal will, by GOD's grace, be found the plain and perfect A. B. C., most necessary for the commons and every subject in his calling to be carefully and diligently musing upon, or exercising himself therein; till, shortly, they may be able in effect to read before their eyes, the most joyful and pleasant British histories (by that Alphabet only deciphered, and so brought to their understanding and knowledge) that ever to this or any kingdom in the whole world else, was known or perceived.

11. Furthermore, how acceptable a thing may this be to the Ragusyes [*Argosies*], Hulks, Caravels, and other foreign rich laden ships, passing within or by any of the sea limits of Her Majesty's royalty; even there to be now in most security where only, heretofore, they have been in most jeopardy: as well by the ravin of the pirate, as the rage of the sea distressing them, for lack of succour, or good and ready pilotage! What great friendship in heart of foreign Prince and subject! And what liberal presents and foreign contributions in hand will duly follow thereof, who cannot imagine?

12. Moreover, such a Petty Navy Royal, *said he,* would be in such stead, as though (a) one [fleet] were appointed to consider and listen to the doings of Ireland; and (b) another to have as good an eye, and ready hand for Scottish dealings; (c) another to intercept or understand all privy conspiracies, by sea to be communicated; and privy aids of men, munition, or money by sea to be transported; to the endamaging of this kingdom, any way intended: (d) another against all sudden foreign attempts: (e) another to oversee the foreign fishermen: (f) another against all pirates haunting our seas: and therewith as well to waft and guard our own merchant fleets as they shall pass and repass between this realm, and wheresoever else they may best be planted for their ordinary marts' keeping; if England may not best serve that turn. And also to defend, help, and direct many of our foreign friends, who must needs pass by or frequent any of those seas, whose principal royalty, undoubtedly, is to the Imperial Crown of these British Islands appropriate.

One such Navy, *said he,* by royal direction, excellently well manned, and to all purposes aptly and plentifully furnished and appointed; and *now, in time of our peace and quiet everywhere, yet beforehand set forth to the foresaid seas* with their charges and commissions (most secretly to be kept from all foes and foreigners) would stand this common wealth in as great stead as four times so many ships would or could do; if, upon the sudden and all at once, we should be forced to deal for removing the foresaid sundry principal matters of annoyance: we being then utterly unready thereto, and the enemy's attempt requiring speedy, and admitting of no successive, defeating.

13. To conclude herein. This Petty Navy Royal undoubtedly will

stand the realm in better stead than the enjoying of four such forts or towns as Calais and Boulogne only could do. For this will be as great strength, and to as good purpose in any coast of England, Ireland or Scotland, between us and the foreign foe, as ever Calais was for that only one place that it is situated in; and will help to enjoy the Royalty and Sovereignty of the Narrow Seas throughout, and of other our seas also, more serviceable than Calais or Boulogne ever did or could do: if all the provisos hereto appertaining be duly observed. Forasmuch as we intend now *peace only preserving,* and no invasion of France or any enemy on that main inhabiting; toward whom by Calais or Boulogne we need to let in our land forces, &c. Much I know may be here said, *Pro et Contra,* in this case: but GOD hath suffered such matters to fall so out; and all to us for the best, if it be so, thankfully construed and duly considered.

For when all foreign Princes, our neighbours, doubtful friends, or undutiful people, subjects or vassals to our Sovereign, perceive such a Petty Navy Royal hovering purposely here and there, ever ready and able to overthrow any of their malicious and subtle secret attempts intended against the weal public of this noble Kingdom in any part or coast thereof: then, every one of them will or may think that, of purpose, that Navy was made out only to prevent them, and none other; and for their destruction, being bewrayed [betrayed] as they would deem. So that not one such foreign enemy would adventure, first, to break out into any notable disorder against us; nor homish subject or wavering vassal, for like respects, durst, then, privily muster to rebellion, or make harmful rodes [inroads] or dangerous riots in any English or Irish Marches.

But such matter as this, I judge you have, or might have heard of, ere now, by worshipful Master DYER; and that abundantly: seeing *Synopsis Reipublicæ Britanicæ,* was, at his request, six years past [*i.e. in* 1570] contrived; as by the methodical author thereof, I understand. Whose policy for the partings, meetings, followings, circuits, &c., of the ships (to the foresaid Petty Navy Royal belonging) with the alterations both of times, places and numbers, &c., is very strange to hear.

So that, in total sum of all the foresaid considerations united in one, it seemeth to be almost a mathematical demonstration, next under the merciful and mighty protection of GOD, for a feasible policy to bring and preserve this victorious British Monarchy in a marvellous security. Whereupon, the revenue of the Crown of England and wealth public will wonderfully increase and flourish; and then, thereupon, sea forces anew to be increased proportionally, &c. And so the Fame, Renown, Estimation, and Love or Fear of this British *Microcosmus,* all the whole and great World over, will be speedily be spread, and surely be settled, &c.

[From Edward Arber (ed.), *An English Garner,* vol. 2, (1879)]

FISHERIES

IT IS most earnestly and carefully to be considered that our herring fishings, [over] against Yarmouth chiefly, have not (so notably, to our great injury and loss and the great and incredible gain of the Low Countries) been traded, but from Thirty-six years ago hitherward. [*This fixes the commencement of the Dutch-herring fishery on the English coasts about* 1540.] In which time, as they have in wealth, and numbers of boats and men, by little and little increased, and are now become very rich, strong, proud, and violent; so, in the race [*course*] of the selfsame time running, the coasts of Norfolk and Suffolk next to those fishing-places adjacent, are decayed in their navy to the number of 140 Sail, and they [of] from threescore to a hundred tons and upwards [each]; besides Crayers and others. Whereupon, besides many other damages thereby sustained publicly, these coasts are not able to trade to Iceland, as in times past they have done; to no little loss yearly to the wealth public of this kingdom.

But the Herring Busses hither yearly restoring out of the Low Countries, under King PHILIP his dominion, are above 500.

Besides 100 or such a thing, of Frenchmen.

The North Seas fishing, within the English limits, are yearly possessed of 300 or 400 Sail of Flemings [*Dutch*]; so accounted.

The Western fishings of Hake and Pilchards are yearly possessed by a great navy of Frenchmen; who yearly do great injuries to our poor countrymen, Her Majesty's faithful subjects.

Strangers also enjoy at their pleasure the Herring fishing of Allonby, Workington, and Whitehaven on the coast of Lancashire.

And in Wales, about Dyfi [*the Dovey*] and Aberystwith, the plentiful Herring fishing is enjoyed by 300 Sail of strangers.

But in Ireland, Baltimore [*near Cape Clear*] is possessed yearly, from July to Michaelmas most commonly, with ' 300 Sail of Spaniards, entering there into the fishing at a Strait [*passage*] not so broad as half the breadth of the Thames [over] against Whitehall. Where, our late good King EDWARD VI.'s most honourable Privy Council was of the mind once to have planted a strong bulwark [*fort*]; for other weighty reasons, as well as His Majesty to be Sovereign Lord of the fishing of Millwin and Cod there.

Black Rock [*? co. Cork*] is yearly fished by 300 or sometimes 400 Sail of Spaniards and Frenchmen.

But to reckon all, I should be too tedious to you; and make my heart to ache for sorrow, &c.

Yet surely I think it necessary to leave to our posterity some remembrance of the places where our rich fishings else are, about Ireland. As at Kinsale, Cork, Carlingford, Saltesses, Dungarven, Youghal, Waterford, La Foy, The Band, Calibeg [*Killibegs*], &c. And all chiefly enjoyed, as securely and freely from us by strangers, as if they were within their own Kings' peculiar sea limits: nay, rather as if those coasts, seas, and bays, &c., were of their private and several purchases. To our unspeakable loss, discredit, and discomfort; and to no small further danger in these perilous times, of most subtle treacheries and fickle fidelity.

Dictum, Sapienti sat esto.

[From Edward Arber (ed.), *An English Garner,* vol. 2 (1879).]

THE BRITISH EMPIRE

I AM not utterly ignorant (said he) of the humours and inclinations of the people of this *Albion,* being (now) the greater portion of THE BRITISH EMPIRE. For through so many conquests as also great immigration hither, there has been made a marvellous mixture of people in sharply contrasted conditions. Yet from year to year, the general disposition of the present inhabitants continually shifts toward this one great imperfection: that is, whilst they know and taste of the best, yet they seldom constantly pursue it.

I mean, for example, in public behaviour *et officiis civilibus:* for there, their civil conversation and industry is in many points not as suited to the dignity of Man as the very Heathens, who did prescribe rules for government. Let CICERO in his Golden Book, *De Officiis,* be the evidence to the contrary against them, particularly in those points, expressed by the heathen orator, which are both greatly agreeable to the most sacred, divine oracles of our God and to the Commonwealth's right excellent prosperity.

I have oftentimes (said he) and in many ways looked into the general state of earthly kingdoms the whole world over (as far as it may yet be known to Christian men commonly), it being a study of no great difficulty, but rather animated by a purpose somewhat similar to that of a perfect Cosmographer: to find himself *Cosmopolites:* a Citizen and Member of the whole, and one and only Mystic City Universal, and so consequently to meditate on the Cosmopolitical Government thereof . . .

And I find (said he) that if this British Monarchy would hitherto have followed up the advantages which it has had continually, it might very well, before this, have surpassed (in a just and godly way) any other particular Monarchy that ever was on Earth since Man's Creation. And that, still serving all such purposes as are most acceptable to God, being of all perfect Commonwealths the most honourable, profitable and comfortable.

(NOTE: The British Monarchy has been capable of the greatest civil felicity that ever was among any other particular Monarchy in the whole world: yea, so incomparably so that it might have contended for the General Monarchy with any that has been, if the requisite policy had been used in due time and constantly followed).

But yet (said he) there is a little lock of LADY OCCASION's hair flickering in the air for our hands to catch hold of, whereby we may, yet once more, (before all be utterly past and gone for ever), discreetly and valiantly recover and enjoy, if not all our Ancient and due appurtenances to this Imperial British Monarchy, yet at the least, some such notable portion thereof as (all circumstances duly and justly appertaining to peace and amity with Foreign Princes being the case) ensures that this may become the most peacable, richest, most puissant and most flourishing Monarchy of all else at this time in Christendom. Peacable (I say) even with the most part of the same respects that good King Edgar had (being but a Saxon); and by such means as he above all in this Empire did put into practice triumphantly. Whereupon his surname was *Pacificus,* most aptly and justly. This peacable King Edgar (about 600 years ago) had in his mind the notion of a great part of the same *Idea* which (from above only and by no Man's advice) has graciously streamed down into my imagination; being (as it becomes me, a subject) careful for the Godly prosperity of this British Empire under our most peacable Queen Elizabeth.

CONCLUSION

SEEING THEN:

No Kingdom in these days has more need of a Petty Navy Royal, to be maintained continually at sea for the reasons enumerated above,

No Kingdom has apter Timber for shipping and still enough of it,

No Kingdom has more and more skilful Shipwrights,

No Kingdom has subjects better able and more willing to contribute towards the sufficient setting forth and maintenance continually at sea of such a Petty Navy Royal,

No Kingdom has a better store of apt and willing men, courageous gentlemen as well as others very manfully disposed to furnish the aforesaid Navy for every kind of purpose,

No Kingdom has better or more havens and harbours (and those all round about it) to succour a Navy from dangers or distress at sea,

No King nor Kingdom has by Nature and Human Industry (to be used) any more lawful and more peacable means (as has been made evident) whereby to surpass all others in Wealth; to become in Strength and Force, INVINCIBLE: and in honourable estimation to be Triumphantly Famous over all and above all others, THAN THIS HAS.

And (to be brief), seeing that no Kingdom is more discreet and more willing to use the opportunity of having any exceeding great public benefit than this British Monarchy is or may be: our hope, then, is upon the uniform, brotherly, willing, and frank consent, in all states of men and people, of this incomparable realm of England, to these godly, politic and commendable means:

1 To preserve Amity and Peace with all Foreign Princes, and
2 To guard this Public State in Security from taking injury of any or by any, fraudulently or forcibly, and to
3 Keep our own hands and hearts from doing or intending injury to any foreigner on Sea or Land.

Our hope is (said he) that upon this Godly Intent, Discreet Covenant and Public Contributory Oblation, the Omnipotent Author of Heavenly Peace will so bend down His merciful and gracious Eyes upon us, and manifestly stretch forth His Almighty Hand to bless, further and prosper the aforesaid Oblation with all the purposes and commodities thereof expected and likely to ensue: that we all may with the Kingly Prophet David — both old and young, rich and poor — most joyfully and triumphantly, IN PERFECT SECURITY, sing Psalm 147:

'O Jerusalem, praise the Lord. Praise thy God. O Syon. For He hath strengthened the bars of thy Gates. And hath blessed thy Children within thee. He hath made all thy Borders PEACE. And with the good Nutriment of wheat, cloth, satisfy thee &c. He hath not done thus to every Nation, else. Praise we all the Lord therefore.

'Amen.'

[Transliteration: GS.]

7

DIARIES
[1577-83]

WHO knew Dee? And what was the nature of his influence in cultured Elizabethan circles? The interested reader is strongly recommended to turn to two admirable pieces of scholarly analysis: *John Dee And The Faerie Queen*[1] by Frances Yates; and *John Dee and the Sydney Circle*[2] by Peter French. Here it is only possible to sketch an outline of the answers.

'Would to God in heaven I had awhile . . . the mysticall and supermetaphysicall philosophy of Doctor Dee,'[3] Gabriel Harvey wrote to Edmund Spenser. And according to Yates, it is extremely likely that Spenser endorsed this view, for in 1580 he was in contact with Dee's pupils, Edmund Dyer and Sir Philip Sydney, who were close friends with one another. Dyer was Dee's constant supporter and disciple, and also a friend of Mary Sydney who, when she married the Earl of Pembroke, kept as her chemist Adrian Gilbert, pupil of Dee, half-brother of Sir Walter Raleigh, and brother of another Dee student, Sir Humphrey Gilbert.

Dee's influence is reflected in the work of Sydney, the favourite nephew of Dee's former pupil the Earl of Leicester, who continued throughout his life to support Dee's projects. Sir Francis Walsingham was another close friend, often inviting Dee to his house at nearby Barn Elms. Sydney eventually married Frances Walsingham who, as Countess of Essex, became godmother to one of Dee's children.

Passages in the *Preface* to *Euclid* clearly demonstrate Dee's keen interest in architecture, poetry, music, and the arts generally. His library must have acted as a magnet, drawing to it all who desired greater learning: 'Dee was the true philosopher of the Elizabethan age,' Yates declares, 'and Spenser as its epic poet, reflected that philosophy.'[4]

The *Diaries* 1577-83 mention other notables with whom Dee was acquainted, principally the Queen herself. There are two selections here. One dates from 1889 and is edited pseudonymously, by 'Hippocrates Junior'.[5] Later we shall be returning to questions raised by the entries for 28 October, 4, 7, 9, 14 November, and 11 and 15 December. Why was Dee 'directed to my voyage' by the Earl of Leicester and Mr Secretary Walsingham? And what was the purpose of the journey to Frankfurt-upon-Oder, where a special messenger was sent to him?

The second extract is edited by James O. Halliwell (1842).[6] Here we can discern Dee's increasing preoccupation with practical occultism, for the 'Barnabus Saul' referred to on 27 January 1582 was Dee's first 'scryer in the spirit vision': and the 'Mr Talbot' mentioned on 9 March is none other than Edward Kelley.

❦

DIARIES
(i)

1577

Jan 16th.   The earl of Leicester, Mr Philip Sydney, Mr Dyer &c. came to my house.

Jan 22nd.   The Earl of Bedford came to my house.

Nov 3rd.   William Rogers of Mortlake, about seven o'clock in the morning, cut his own throat.

Nov 22nd.   I road to Windsor to the Queen's Majesty.

Nov 15th.   I spoke with the Queen *hora quinta*.

Nov 28th.   I spoke with the Queen *hora quinta* . . . I declared to the Queen her title to Greenland, Estetiland and Friesland.

1578

Sept 1st.   I came from Cheyham, Sept 6th. Elen Lyne, my maiden, departed from this life immediately after mid-day, when she had lain sick a month less one day.

Sept 12th.   Jane Gaele came into my service, and she must have four nobles by the year, 26s 8d.

Sept 25th.   Her Majesty came to Richmond from Greenwich.

Sept 26th.   The first rain that came for many a day; all pasture about us was withered; rain in the afternoon like April showers.

Oct 8th.   The Queen's Majesty had conference with me at Richmond, *inter 9 et 11*.

Oct 16th.   Dr Bayly conferred of the Queen's disease.

Oct 22nd.   Jane Fromonds went to the court at Richmond.

Oct 25th.	A fit from 9 afternoon to 1 after midnight.
Oct 28th.	The Earl of Leicester and Sir Francis Walsingham determined my going over for the Queen's Majesty.
Nov 4th.	I was directed to my voyage by the Earl of Leicester and Mr Secretary Walsingham *hora nona*.
Nov 7th.	I came to Gravesend.
Nov 9th.	I went from Lee to sea.
Nov 14th.	I came to Hamburg *hora tertia*.
Dec 11th.	To Frankfurt-upon-Oder.
Dec 15th.	News of Turnifer's coming, *hora octava mane*, by a special messenger.

1579

June 15th.	My mother surrendered Mortlake houses and land and had state given in *plena curia ad terminam vitae*, and to me also the reversion delivered *per virgam*, and to my wife Jane by me, and after to my heirs and assignees for ever, to understand, Mr Bullock and Mr Taylor, surveyor at Wimbledon, under the tree by the church.
Oct 31st.	Paid xxs fine for me and Jane my wife to the Lord of Wimbledon (the Queen) by goodman Burton of Putney, for the surrender taken of my mother of all she has in Mortlake to Jane and me and then to my heirs and assignees, &c.

1580

June 7th.	Mr Skydmoor and his wife lay at my house and Mr Skydmoor's daughter, and the Queen's dwarf Mrs Tomasin.
Sept 6th.	The Queen's Majesty came to Richmond.
Sept 10th.	Sir Humphrey Gilbert granted me my request to him, made by letter, for the royalties of discovery all to the North above the parallel of the 50th degree of latitude, in the presence of Stoner, Sir John Gilbert, his servant or retainer; and thereupon took me by the hand with faithful promises in his lodging of John Cook's house in Wichcross Street, where we dined only us three together, being Saturday.
Sept 17th.	The Queen's Majesty came from Richmond in her coach, the higher way of Mortlake field, and when she came right

against the church, she turned down toward my house; and when she was against my garden in the field, she stood there a good while, and then came into the street at the great gate of the field, where she espied me at my door making obeisance to her Majesty; she beckoned her hand for me; I came to her coach side, she very speedily pulled off her glove and gave me her hand to kiss; and to be short, asked me to resort to her court and to give her to wete when I came there; hor. 6¼ *a meridie.*

Oct 3rd.
On Monday, at 11 o' clock before noon, I delivered my two rolls of the Queen's Majesty's title to herself in the garden at Richmond, who appointed after dinner to hear further of the matter. Therefore between one and two in the afternoon, I was sent for into her highness' Privy Chamber, where the Lord Treasurer also was, who, having the matter slightly then in consultation, did seem to doubt much that I had or could make the argument probable for her highness' title as I pretended. Wherapon I was to declare to his honour more plainly and at his leisure what I had said and could say therein, which I did on Tuesday and Wednesday following, at his chamber, where he used me very honourably on his behalf.

Oct 10th.
The Queen's Majesty, to my great comfort (*hora quinta*) came with her train from the court, and at my door graciously calling me to her, on horseback, exhorted me briefly to take my mother's death patiently, and withal told me that the Lord Treasurer had greatly commended my doings for her title, which he had to examine, which title in two rolls he had brought home two hours before; she remembered also how at my wife's death it was her fortune likewise to call upon me. At 4 o' clock in the morning, my mother Jane Dee died at Mortlake; she made a godly end; God be praised therefore! She was 77 years old.

Dec 1st.
The Queen lying at Richmond.

Dec 6th.
The Queen removed from Richmond.

(ii)

Jan. 16th, Mistris Harbert cam to Essexe. Jan. 17th, Randal Hatton

cam home from Samuel's father at Stratton Audley. Jan. 22nd, Arthur
Dee and Mary Herbert, they being but 3 yere old the eldest, did make
as it wer a shew of childish marriage, of calling ech other husband
and wife. Jan. 22, 23rd. The first day Mary Herbert cam to her father's
hous at Mortlak, and the second day she cam to her father's howse
at Estshene. Jan. 23rd, my wife went to nurse Garret and payd her
for this month ending the 26 day. Jan. 27th, Barnabas Sawl his brother
cam. Feb. 12th, abowt 9 of the clok, Barnabas Saul and his brother
Edward went homward from Mortlak: Saul his inditement being by
law fownd insufficient at Westminster Hall: Mr. Serjeant Walmesley,
Mr. Owen and Mr. Hyde, his laywers at the bar for the matter, and
Mr. Ive, the clerk of the Crown Office, favouring the other. Feb. 20th,
Mr. Bigs of Stentley by Huntingdon and John Littlechild cam to me.
I receyved a letter from Barnabas Saul. Feb. 21st, Mr. Skullthorp rod
toward Barnabas. Feb. 25th, Mr. Skulthorp cam home. Payd nurse Garret
for Katharin tyll Fryday the 23 day, vjs. then somethyng due to nurse
for iij. pownd of candell and 4 pownd of sope.

March 1st, Mr. Clerkson browght Magnus to me at Mortlak, and
so went that day agayn. March 6th, Barnabas Saul cam this day agayn
abowt one of the clok and went to London the same afternone. He
confessed that he neyther hard or saw any spirituall creature any more.
March 8th, Mr. Clerkson and his frende cam to my howse. Barnabas
went home agayn abowt 3 or 2 clok, he lay not at my howse now;
he went, I say, on Thursday, with Mr. Clerkson. March 8th, coelum
ardere et instar sanguinis in diversis partibus rubere visum est circa
horam nonam noctis, maxime versus septentrionalem et occidentalem
partem: sed ultra capita nostra versus austrum frequenter miles quasi
sanguineus. March 9th, Fryday at dynner tyme Mr. Clerkson and Mr.
Talbot declared a great deale of Barnabas nowghty dealing toward me,
as in telling Mr. Clerkson ill things of me that I should mak his frend,
as that he was wery of me, that I wold so flatter his frende the lerned
man what I wold borow him of him. But his frend told me, before
my wife and Mr. Clerkson, that a spirituall creature told him that
Barnabas had censured both Mr. Clerkson and me. The injuries which
this Barnabas had done me diverse wayes were very great. March 22nd,
Mr. Talbot went to London, to take his jornay.

April 16th, Nurse Garet had her 6s. for her month ending on
the 20th day. April 22nd, a goodly showr of rayn this morning early.
May 4th, Mr. Talbot went. May 13th, Jane rod to Cheyham. May

15th, nocte circa nonam cometa apparuit in septentrione versus occidentem aliquantulum; cauda versus astrum tendente valde magna, et stella ipsa vix sex gradus super horizontem. May 20th, Robertus Gardinerus Salopiensis laetum mihi attulit nuncium de materia lapidis, divinitus sibi revelatus de qua . . . . May 23rd, Robert Gardener declared unto me hora 4½ a certeyn great philosophicall secret, as he had termed it, of a spirituall creatuer, and was this day willed to come to me and declare it, which was solemnly done, and with common prayer. May 28th, Mr. Eton of London cam with his son-in-law Mr. Edward Bragden, as concerning Upton parsonage, to have me to resign or let it unto his said son-in-law, whom I promised to let understand whenever myself wold consent to forego it. June 9th, I writ to the Archbishop of Canterbury a letter in Latin: Mr. Doctor Awbrey did carry it. June 14th, Morryce Kyffin did viset me.

8

SPIRITUAL DIARIES
[1583-7]

Glendower: I can call spirits from the vasty deep.
Hotspur: Why so can I, so can any man, but will they come when you do
call for them?

Shakespeare; King Henry IV Part I, III. i.

EARLIER, we noted the influence upon Renaissance occult philosophy of
Henry Cornelius Agrippa, who urged that the universe be conceived as
consisting of three worlds: the world of Elemental or Terrestrial Nature, which
was the province of the physical sciences; the Celestial World of the stars,
which could be understood and manipulated by the study and practice of
Astrology and Alchemy; and the Supercelestial World, which could be
apprehended by numerical operations and the conjuring of the angels themselves.
It was this Supercelestial World that Dee now set out to explore.

Theory was not enough for Agrippa: nor was it enough for Dee. Hence
his angel magic, in which his desire for further knowledge and wisdom was
powered also by the philosophy of his age and by motives which were both
religious and scientific. They were religious in that Dee sincerely believed
himself to be dealing with emissaries of God, and so consistently displayed
an attitude of Christian piety. They were scientific in that Dee was investigating
the question: is there intelligent life in other dimensions? He believed that
there is and that Man can improve his state by communion with angels.

However, in order for Dee to make contact with these angels, he needed
a medium capable of seeing and hearing them. In his first recorded conference
of 22 December 1581, the seer was Barnabas Saul, who later denied that
he had seen anything and proceeded to slander Dee who wrote: 'The injuries
which this Barnabas had done me diverse wayes were very great.'[1] Saul was
succeeded by Edward Kelley.

Few commentators have a good word to say for Kelley. Little enough
is known about his life. Born at Worcester on 1 August 1555, he attended
Oxford as 'Edward Talbot' for a time but left under a cloud. A few years

later, he was pilloried in Lancaster for forgery. Subsequently he was the secretary of the mathematician and Hermetic scholar Thomas Allen, though this period of service was brief and, as we have seen, he presented himself at Dee's house at Mortlake on 10 March 1582. According to Dr Thomas Head: 'The portrait that emerges from the record of his sittings with Dee is of a highly ambiguous personality, wary and mistrustful, unstable and picric, prone on the one hand to terrifying fits of anger accompanied by physical violence, and on the other hand to sudden spiritual conversions from which he promptly relapsed.'[2]

The contrast between the life and character of Dee and that of Kelley must have been a source of perennial fascination for the two men. Dee was attracted also by Kelley's claim that he was a practical alchemist; nor could Dee pursue angel magic successfully without the mediumship of someone like Kelley. The initial results were exciting for they included a remarkable prophecy, which is the subject-matter of our first extract:

<hr/>

from Liber Logaeth (Sloane MS 3189)

5 May 1583

Dee: As concerning the Vision which yesternight was presented (unlooked for) to the sight of EK as he sat at supper with me, in my hall, I meane the *appering of the very sea,* and many ships thereon, and the cutting of the hed of a woman, by a tall blak man, what are we to imagin thereof?

Ur[iel]: The one did signifie the provision of forrayn powres against the welfare of this land: which they shall shortly put into practice. The other, the death of the Queene of Scotts. It is not long unto it.

Mary Queen of Scots was beheaded in 1587: the Armada sailed in 1588.

How was such communion attained? The initial preparations were simple, as Dr Head notes; they consisted

merely of setting up the shewstone[3] or crystal on the table of practice and of a short prayer spoken by the Doctor. The result was that Kelley received, on the first day, a vision of the angel Uriel, who revealed his secret signature and issued preliminary directions for the construction of two magical talismans: (1) The Sigillum Dei Aemeth, a pentacle nine inches in diameter, to be made of purified wax; and (2) the Tabula Sancta, a table to be made of sweet wood two cubits high and two cubits square, on which a large rectangular seal containing twelve Enochian letters was to be surrounded with seven circular seals attributed to the planetary powers. The two talismans — which were in fact the first two Enochian documents — were to be employed together, the pentacle being placed on the Holy Table while in use. [4]

Now the complexity of events increased. On 14 March, a spirit purporting to be the angel Michael gave instructions for making a magic ring of gold bearing a seal said to be identical to that 'wherewith all miracles and divine works and wonders were wrought by Solomon'. On 20 March, the angel Uriel dictated a square of forty-nine characters containing seven angelic names, and a day later, a second square was dictated. And it was shortly after this that Kelley started to produce an enormous amount of material concerning an angelical or 'Enochian' language.

As Head writes:

The Enochian alphabet appeared first — twenty-one characters, somewhat like Ethiopic in styling though not in formation, and written like all Semitic languages from right to left. This was followed by a book containing almost one hundred squares, many of them as large as 2,401 characters (49 × 49), whose dictation became the principal business of all the sittings for nearly fourteen months. And the material continued to pile up, page after page, book after book, until the final parting between Dee and Kelley in 1589. [5]

The matter may be studied in the original manuscripts and also in the large selection printed as *A True & Faithful Relation of what passed for many Yeers Between Dr John Dee . . . and Some Spirits,* edited by Meric Casaubon, published in London, 1659, and reprinted in 1976. [6]

The following extract forms the opening of the Casaubon text but reminds one more of a spiritualists' seance than of the conjuring of angels.

LIBER MYSTERIORUM (& SANCTI) PARALLELUS NOVALISQUE

Leyden, May 28 1583

As I and E.K. sat discoursing of the noble Pole, Albertus Lasky, of his great honour obtained here with us, of his great popularity with all kinds of people, those that see him or hear of him, and again of how much I was beholden to God that his heart should so fervently favour me, and that he does strive so much to suppress and confound the malice and envy of my countrymen against me, winning for my better credit, or recovering it, that it may do God better service hereafter, &c . . .

Suddenly, there seemed to come out of my Oratory *a Spiritual Creature,* like a pretty girl of 7 or 9 years of age, attired on her head with her hair rolled up before, and hanging down very long behind, with a gown of Sey . . . changeable green and red, and with a train she seemed to play up and down . . . like, and seemed to go in and out behind my books, lying on the biggest heaps . . . and as she should ever go between them, the books seemed to give place sufficiently . . . one heap from the other, while she passed between them. And so I considered, and . . . the divers reports which E.K. made to me of this pretty maiden, and . . .

I said Whose maiden are you?

She: *Whose man are you?*

I am the servant of God both by my bound duty, and also (I hope) by His Adoption.

A VOICE: *You shall be beaten if you tell.*

She: *Am I not a fine Maiden? Give me leave to play in your house, my Mother told me she would come and dwell here.*

She went up and down with the most lively gestures of a young girl, playing by herself, and divers times another spake to her from the corner of my study by a great Perspective glass, but none was seen beside herself.

She: *Shall I? I will.* (Now she seemed to answer one in the foresaid corner of the study). *I pray you, let me tarry a little.* (Speaking to one in the foresaid corner).

Tell me who you are?

I pray you let me play with you a little, and I will tell you who I am.

In the name of Jesus then tell me.

I rejoice in the name of Jesus, and I am a poor little Maiden, Madini,[7] *I am the last but one of my Mother's children, I have little Baby-children at home.*

Where is your home?

I dare not tell you where I dwell, I shall be beaten.

You shall not be beaten for telling the truth to them that love the truth; to the eternal truth all Creatures must be obedient.

I warrant you I will be obedient. My Sisters say they must all come and dwell with you.

In the next extract, Dee, Kelley and their respective wives are in Cracow, communicating not only with angels but also with their sometime patron, King Stephen of Poland, as a result of the intercession of their friend 'the noble Pole, Albertus Lasky'.

The record makes for wearisome reading but is printed here to show the complexity of the Enochian System, the tedium which must have accompanied the transmissions and of which the purported angel warns, and the persistence of Dee and Kelley.

1584

Friday morning, Hora 8½, April 13, Cracow

Not long after my Invitation, *Nalvage* appeared, *Nutu Dei.*

NAL: *Our peace, which is* Triumphing patience, *and glory be amongst you.* Amen.

NAL: *It may be said, can there be patience in the Angels, which are exalted above the aire? For, such as were of error have their reward: Yea, forsooth my dear brethren. For there is* a continual fight between us and Satan, *wherein we vanquish by patience. This is not spoken without a cause. For as the Devil is the father of Carping so doth he utterly infect the* Seer's *imagination,* mingling unperfect forms with my utterance. *Water is not received without air, neither the word of God without blasphemous insinuation. The Son of God never did convert all, neither did all that did hear Him believe Him. Therefore,* where the power of God is, is also Satan: *Lo, I speak not this without a cause, for I have answered thy infection.*

E.K. had thought that Angels had not occasion of any patience, and so was his thought answered.

NAL: *I find the Soul of man hath no portion* in this first Table. *It is the Image of the Son of God, in the bosom of His father, before all the worlds. It comprehendeth His incarnation, and return to judgement: which He Himself,* in flesh, *knoweth not; all the rest are of understanding.* The exact Centre excepted.

A (*Two thousand and fourteen, in the sixth Table, is*) D.
86. 7003. *In the thirteenth Table, is I.*

A *In the 21st Table. 11406 downward.*

I *In the last Table, one less then Number. A word, Jaida you shall understand, what that word is before the Sun go down. Jaida is the last word of the call.*
85. *H 49. ascending T 49. descending, A 909. directly, O simply. H 2029. directly, call it* Hoath.

225. *From the low angle on the* right side. Continuing in the same and next square.

D *225 (The same number repeated).*

A *In the thirteenth Table, 740, ascending in his square.*

M *The 30th Table, 13025. from the low angle in the left side.*
84 . . . *In the square ascending.*

Call it Mad.

O *The 7th Table, 99. ascending.*

C *The 19th. descending 409.*

O *The . . . 1. from the upper right angle, crossing to the nether left and so ascending 1003.*

83. N. *The 31st from the Centre to the upper right angle, and so descending 5009.*

Call it Noco.

Be patient, for I told you it would be tedious.

O *The 39th, from the Centre descending, or the left hand, 9073.*

D *The 41st from the Centre ascending, and so to the right upper Angle, 27004.*

R *The 43rd. from the upper left Angle to the right, and so still in the Circumference, 34006.*

I *The 47th, ascending, 72000.*

82. *In the same Table descending the last.*

Call it Zirdo.

P *The 6th. ascending 109.*

A *The 9th. ascending 405.*

81. L. *The 11th, descending 603 . . .*

Call it Lap.

Here, he stroke the Table on Saturday action following at my reading over of it backward.

E *The 6th from the right Angle uppermost to the left, 700.*

G *The 13th descending, 2000.*

R *The 17th from the Centre downward, 11004.*

80. O, *The 32nd, descending from the right Angle to the Centre, 32000.*

Z. *47th, 194000, descending. Call it* Zorge *(Of one syllable).*

A *19th, from the left corner descending, 17200.*

79. A *24th, from the Centre ascending to the left Angle, 25000.*

Q *The same Table ascending, 33000.*

Call it Q A A *(Three syllables with accent on the last A)*

E *The second Table, 112 ascending.*

L *The . . .; descending 504.*

C *The 19th Table descending 1013 (That C is called C Minor).*

I *The 13th descending, 2005.*

C *The 14th descending, 2907. Call it* Cicle.

E.K.: Now he is kneeling, and praying with his Rod up.

Prague is the scene of the following extract, for it was there that Dee and Kelley sought and for a time enjoyed the patronage of the Emperor Rudolf II.

The style, tone and content are probably best enjoyed if the reader bears in mind Dee's conviction that the communicator is a messenger from God Almighty.

1585

6 August: Prague

URIEL: Thy mouth, O Lord, is a two edged sword, thy judgements are perpetual and everlasting, thy words are the spirit of truth and understanding, thy garments most pure and smelling incense; thy Seat without end and triumphing. Who is like unto Thee amongst the heavens, or who hath known thy beauty? Great art thou in thy holy ones, and mighty in thy word amongst the Sons of men. Thy Testament is holy and undefiled, the glory of thy Seat, and the health of thy Sons. Thy anointed is sacrificed and hath brought health unto the faithful and unto the sons of Abraham. Thy Spirit is everlasting, and the oil of comfort. The Heavens (therefore) gather themselves together, with Hallelujah to bear witness of thy great indignation and fury prepared for the Earth, which hath risen up with the Kings of the Earth, and hath put on the Wedding Garments: saying with her self, I am a Queen, I am the daughter of felicity. Remember all ye that are drunken with my pleasure, the Character I have given you, and prepare yourselves to contend with the Highest, set yourselves against Him, as against the anointed, for you are become the Children of a strong Champion, whose Sun shall garnish you with the Name of a kingdom and shall pour wonders amongst you, from the stars, which shall put the Sun the Steward of his Waggon and the Moon the handmaid of his servants. But, O God, she is a Liar and the firebrand of destruction. For behold, thou art mighty, and shall triumph and shall be a Conqueror forever.

E.K.: Now the Stone is full of white smoke.

A PAUSE

E.K.: The smoke is gone, and here stands one over him in the air with a Book, whose nether parts are in a cloud of fire, with his hair sparse, his arms naked, the Book is in his right hand, a four square Book, with a red, fiery cover, and the leaves be white on the edge; it hath 7 seals upon it, as if the clasps were sealed with 7 golden seals. And there are letters upon the Seals, the first E.M.E.T.T.A.V.

THE ANGEL WITH THE BOOK: Take this Book, ut veritas Luce magis clareseat, Et Lux, veritate fiat valida. Data est enim tibi puestas, dandi & aperiendi hunc liorum mundo & munaiis.

URIEL: Gloria tibi Rex coelu & terra qui fuistis & venturus, es hinc enim judiciu meretricis.

E.K.: Now URIEL takes the Book, kneeling upon both his knees.

URIEL: Rejoice O you sons of men, lift up your hearts unto heaven for the secrets of God are opened: and His Word let out of Prison. Rejoice, O you sons of God, for the Spirit of Truth and Understanding is amongst you. Rejoice O you that are of the Sanctuary, for yu shall be full of wisdom and understanding. Rejoice O thou the House of Jacob, *for thy vilification is at an end,* and thy vilification is beginning. The four winds shall gather thee together, and thou shalt build up the trodden wall: the bridegrooms shall dwell with thee. And lo, behold, the Lord hath sworn, and wickedness shall not enter into thee, neither shall the Spirit of the Highest go from thee, but thy father's bones shall have rest: and thou shalt live eternally.

The blood of the Innocents shall be washed away from thee, and thou shalt do penance for many days. Then shall the Lamb stand in the midst of thy streets, O Jerusalem: and shall give Statutes to thy people and inhabitants. All Nations shall come unto the House of David: The Mothers shall teach their infants, saying: Truth hath prevailed, and the Name of the Lord shall be the Watchman of thee, O City.

E.K.: Now all is full of a white cloud.

URIEL: Silence unto me, and rest unto you for a season.

E.K.: All is disappeared, and the Stone seems clear.

ACTIONIS PUCCIANAE POSTERIOR PARS

Legi praemissa Latine ipsi Fr. Pucci, & pauca locutus sum de regibus & aliis qui haec putant esse nostras imposturas, & a nobis haec mala ratione tractari, &c.

E.K.: He is here again. He sits in a chair of Crystal, with his Book in his lap and the *measuring rod* in his right hand, and the glass vial in his left hand.

URIEL: Seeing that power is given unto me, and that truth is added unto my Ministry, and I am become full of light and truth, I will open your eyes, and I will speak unto you the truth that you may shake off the lumpiness of your darkness, and profound ignorance: and walk in truth with your fathers.

Give ear (therefore) diligently unto my voice: and imbibe my sayings, within the liquor of your hearts, that the sap of your understanding may receive strength, and that you may flourish with acceptable Truth, as the chosen servants and Ministers of the Highest.

Totus mundus in maligno positres est, and is become the open shop of Satan, to deceive the Merchants of the Earth with all abomination. But what, are you the Peddlers of such wares of the Carriers abroad of lies and false doctrine? Do you think it is a small matter to tie the sense of God's Scriptures and mysteries unto the sense and snatching of your Imagination? Do you count it nothing to fit in judgement against the Spirit of God: leaving Him no place, but at your limitation? Is it lawful before the Son of God, to spend the whole days, yea, many years, with the Sons of Satan, the lying imps and deceivers of the World? Are you so far entered into the shop of abomination, that you point unto the *Son of God the time* of his coming, the descending of his Prophets, and the time wherein he shall visit the Earth? MOSES durst not speak, but from the Lord's mouth: The Prophets expounded not the Law, but the voice of the Lord. The Son of God spake not his own words, in that he was flesh, but the words of his Father; His Disciples taught not, but through the Holy Ghost. Dare you (therefore) presume to teach, and open the secret Chamber of the Highest, being not called?

Tell me, have you left your Merchandise and the counting of your money deceitfully gotten, to become Teachers of the Word of God? Are you not ashamed to teach before you understand? yes, are you not ashamed to lead away, where you cannot bring home? Hypocrites you are, and void of the Holy Ghost, liars you are become and the enemies of Christ, and his holy Spirit.

Peradventure you will say, in reading the Scriptures we understand. But tell me, by what spirit you understand them: what Angel hath appeared unto you? or of which of the Heavens have you been instructed? It may be you will say of the Holy Ghost, O thou fool, and of little understanding! Dost thou not understand that the Holy Ghost is the Schoolmaster of the Church, of the whole Flock and Congregation of Christ? If he be the School-master (therefore) over a multitude, it followeth then, that one doctrine taught by the Holy Ghost is alessoa or an understanding of a multitude. But what multitude are of thine understanding or of what Congregation art thou? Wilt thou say, thou art scattered? Thou speakest falsely, thou art a renegade. But behold, I teach thee, and thy error is before thy face.

Whosoever doth understand the Scriptures must seek to understand them by Ordinance and spiritual tradition. But of what spiritual tradition understandeth thou? or by what Ordinance are the Scriptures opened unto thee? Thou wilt say thou art informed by the Holy Fathers, and by the same Spirit that they taught, by the same spirit thou understandeth. Thou sayest so, but thou dost not so. Which of thy forefathers *hath tied reason to the word of God?* or the understanding of the Scriptures to the Discipline of the Heathen? I say unto thee, that thy forefathers were dear unto Christ, were partakers of the heavenly visions and celestial comforts, which visions and celestial comforts did not teach unto them a new exposition of the Scriptures, but did confirm and give light unto the mysteries of the Holy Ghost spoken by the Apostles, the ground-layers and founders of the Church. Whatsoever, therefore, thou learnest of thy forefathers, thou learnest of the Apostles, and whatsoever thou learnest of the Apostles, thou hast by the Holy Ghost. But if thou expound the Fathers after thy sense & not after the sense of the Apostles, thou hast not the Holy Ghost but the spirit of lying. Therefore humble thyself and fall down before the Lord. Lay reason aside and cleave unto him. Seek to understand His word according to His Holy Spirit. Which Holy Spirit thou must needs find, and shalt find in *a visible Church, even unto the end.*

I will plainly say unto thee (That Truth may appear mightily in light) Whosoever is contrary unto the will of God, which is delivered unto his Church, taught by His Apostles, nourished by the Holy Ghost, delivered unto the world, and by *Peter* brought to *Rome* by him, there taught by his Successors, held, and maintained, is contrary to God and to His Truth.

Luther hath his reward.

Calvin his reward.

The rest, all that have erred, and wilfully run astray, separating themselves from the Church and Congregation of Christ obstinately,

and through the instigation of their father the Devil, have their reward.
Against whom the Son of God shall pronounce judgement . . .

The following is taken from *An Unknown Chapter in the Life of John Dee*,
edited and commented upon by C. H. Josten and published in *Journal of
the Warburg and Courtauld Institutes* XVIII (1965), 223-57.

An altogether extraordinary chain of events is about to be described,
taking place in Prague just after midday on 10 April 1586. The alleged angel
commands Dee to gather 28 volumes, including one which contains 48
individual books 'most mystical and more valuable than all things in the
whole world might be accounted'. The purpose of this exercise is to burn them.

These instructions are perplexing enough; but the sequel described by
Josten is wholly astonishing.

Whatever I may command you today, take care you do it.
Place the books which lie here also inside the little bag.'

Δ: E[dward] K[elly] put all those 28 books (or primary volumes)
in the same black bag and closed it (tightening the braces as usual).

The voice: 'Now you, Kelly, will rise and you will remove the
stones from the mouth of that furnace, and where those stones now
are you will place this. When it is done, return.'

E[dward] K[elly] rose from his chair and removed from the mouth
of the furnace 4 or 5 fire-baked lateral facings (by which the warmth
of the fire was the more perfectly kept within the furnace, and by
which the inconvenience of cold air and wind coming in was prevented).
He placed the little bag inside the mouth of the burning furnace, in
the very place where previously the aforesaid fire-baked stones had been,
and so returned to his chair.

The voice: 'Now rise and bring hither what remains. Do not conceal
anything.'

Δ: E[dward] K[elly] went and fetched some manuscript quires
consisting each of four sheets (folded in 8°). The writing seemed very
old and was in a larger character than our common script. He had
cut them out of the volume which had been entrusted to his custody
(viz. from the last mentioned book, not from [any of] the former).
When he had brought them and had again sat down with us, the
Lord spoke thus to him:

The voice: 'Rise and throw [them] into the furnace.'

Δ: '(O Lord) dost Thou wish them to be put into the very fire
or on top of the little bag, o God. He is willing (o Lord) to be obedient
to Thy will.'

The voice: 'Rise, Pucci, join him, and see to it that he puts them into the very fire and, besides, that he also thrusts the little bag and the books after them.

You will not withdraw until the fire entirely penetrates them.

Do I not resuscitate the dead?

Go then, and have faith.'

[*In marg: The holocaust of all that which, from the beginning of the world, had been most precious.*] Δ: They arose and went to the furnace. First they threw into the burning fire those quires consisting of four sheets. Then, boldly and briskly, they pushed the little bag into that fire. (While they were thus occupied at the mouth of the furnace,) I raised my voice to God on bended knees and rendered thanks to our God with great joy, gladness, and exultation. I prayed that He might augment and confirm our faith; that we might in no way have doubts about the most generous promises He had previously made to us; and that, to the honour and glory of His name, He might make us most certain witnesses and servants of His wisdom, patience, and goodness. Meanwhile, they applied themselves eagerly and gravely to the completion of the holocaust. They threw into the furnace very light and dry pieces of firewood and very fine chips of timbers and beams in great quantity, stirred the heap of books with a small staff and an iron spit, lifted it, and laid it open, so that the fire might the more easily consume all and convert into ashes and embers whatever would be combustible. And when they had so busied themselves for almost a quarter of an hour, E[dward] K[elly] heard a voice saying to him:

The voice: 'Tell Pucci now that he shall enter (viz. into the oratory).'

[Δ:] Pucci joined me in the oratory and, on bended knees, he poured out with me prayers of thanksgiving, etc.

As soon as Pucci had entered, E[dward] K[elly], who was standing before the mouth of the furnace (now fully ablaze), signified to us in a loud voice that he was seeing the shape of a man (only from the midriff upwards) walking, as it were, hither and thither among the flames, but that his face did not appear to him.

'It seems (said E[dward] K[elly]) that, with his right hand he is gathering, or plucking off, something from the tops of the flames. Now I can see clearly that he is recovering from the flame the leaves of books, leaf after leaf.

Now he seems to have put together a whole book.'

And (after a short while) he said: 'And now he seems to be holding a second book in his hand, yet I do not see where he might have laid away the first.

Now I see that he has collected yet another book among the tops of the flames.

And now it seems to me that, with his fingers, he is shaping from the tops of the flames a small box, with the powder [inside].

He goes on collecting leaves from the tops of the flames.

Now he seems to have [brought together] two further books; *but I cannot see at all where all these things go.'*

'Now indeed that man has disappeared very suddenly,' said E[dward] K[elly].

When all things had been consumed by the fire (that could be consumed) and when this vision had come to an end, E[dward] K[elly] came towards us and said: 'Come, Pucci, see and judge.' He went, considered, and came back. 'It appears to me', he said, 'that nothing is left, except ashes and stones, very few pieces of live coal, a few embers of leaves of paper, the froth of the fire deposited on the sides of the furnace.'

The voice: 'Now you shall add the other things.'

Δ: 'What other things shall I add, o Lord?'

The voice: 'Books . . . Whatever there remains on paper, you shall burn in the same manner as you have burned the former.

The other things you may keep, as I shall teach and instruct you before you leave.'

Δ: 'Is it Thy wish that we burn also this rolled-up and sealed fascicle (which, by Thy command, we took from Fr[ancesco] Pucci)?'

The voice: 'No.'

Δ: 'Shall I add this writing (a small part of the book of Enoch which Thou hast given to me)? Dost Thou want this document also to be burned?'

The voice: 'You shall not burn it.'

Δ: Then I set out on the table all remaining loose and unbound papers and sheets, a great many (all received by divine dictation), and did not let one sheet remain that had any writing on it. They [*above the line:* E[dward] K[elly] and Fr[ancesco] P[ucci]] took all those charts and papers and threw them into the glowing furnace, and very quickly they were ablaze and were burned by the fire ∴ Then they reported this to me, whereupon I said with my whole heart:

Δ: 'Thus I [offer to Thee,] our Almighty, Sempiternal, and Living God, Who art our protector and liberator, this welcome and acceptable sacrifice of our obedience. Amen.'

Δ: Immediately upon this, the divine voice spoke the following words, which were received by E[dward] K[elly] (our seer) and [by him] expressed to me:

The voice: 'Behold, I swear by Myself that not one letter will perish of whatsoever has been committed to the fire. And as I have the power to rouse up glory for Me from nothing, so I have power to collect together what has been brought forth by Me. Therefore, when later on, the tyranny of those men ceases, collect yourselves and make a prayer before Me, invoking the name of the Father in the name of Jesus, His Son. And be aware that, as *these things were put into the fire, in the same way you will receive them again.* [*In marg:* A restitution of the things burned and committed to the fire is promised.] And not one letter of that which I have spoken will perish.

Yet, because they assert that you have intercourse with the enemy of mankind, My visitation upon you has become a cause of offence to them.

The more so, because the mysteries of Heaven cannot be discovered but to those whom I choose and whom (by My vocation) I put apart from the others.

And also, because they will not understand, but will be buried in ignorance (for no one receives from the Father unless he have the merit of faith, nor is anyone approved but he to whom justification has been granted).

For that reason I have prostrated them in the midst of their fancies.

At about 1.30 p.m. on 29 April 1586, Kelly noticed from the gallery of his chamber the head-gardener, and foreman of the workmen, of a Mr Carpio as he seemed to be pruning some trees in Mr Carpio's vineyard; 'at length he approached under the wall by E. K. and holding his face away-ward he said unto him, *Quaeso dicas Domino Doctori quod veniat ad me.* And so went away as it were cutting here and there the trees very handsomely, and at length over the Cherry-trees by the house on the Rock in the Garden he seemed to mount up in a great piller of fire.' Mrs Kelly, whom her husband sent down into the garden, did not see anybody there. Dee and Kelly then descended into the garden together, but they, too, could not find the strange gardener anywhere, who, Kelly presumed, must be 'some wicked spirit'. After a thorough search, Dee suddenly noticed from a distance 'a faire white paper lying tossed to and fro in the wind' under an almond tree, and then, to his great joy, he found lying there three of the books 'which were so diligently burnt the tenth day of April last', namely:

(1) the Book of Enoch;
(2) the '48 Claves Angelicae';
(3) the 'Liber Scientiae terrestris auxilii & victoriae'. (The original, in Dee's handwriting, is now MS. Sloane 3191, fols. 14-31v; a copy, written by Ashmole, is preserved in MS. Sloane 3678, fols. 14-31v.)

The books showed no sign of ever having been in, or near, the fire (Casaubon, p. 418).

Half an hour later, while Dee and Kelly were sitting under the almond tree, 'the self-same Gardiner like person' appeared again (this time perhaps also to Dee). His face was again averted. He bade Kelly follow him, and Dee remained sitting still to await his return. The 'gardener's' feet did not appear to touch the ground and, as he went before Kelly, all the doors seemed to open before him. He conducted Kelly to the mouth of the furnace in the oratory at the top of the tower where all the books and papers had been burnt. 'And coming thither, there the spiritual Creature did seem to set one of his feet on the post on the right hand without the furnace mouth, and with the other to step to the furnace mouth, and so to reach into the furnace (the bricks being not plucked away which stopped the mouth of the furnace, all saving one brick thick) and as he had reached into the furnace there appeared a great light, as if there had been a window in the back of the furnace, and also to E. K. the hole which was no greater then the thickness of a brick unstopped, did seeme now more then three or four brick thickness wide, and so over his shoulder backward he did reach to E. K. all the rest of the standing Books, excepting the Book out of which the last Action was cut, and Fr. Pucci his Recantation, also to E. K. appeared in the furnace all the rest of the papers which were not as then delivered out. That being done, he bade E. K. go, and said he should have the rest afterward. He went before in a little fiery cloud, and E. K. followed with the Books under his arm all along the Gallery, and came down the stairs by Fr. Pucci his Chamber door, and then his guide left E. K. and

he brought me the Books unto my place under the Almond-tree'
(Casaubon, pp. 418-19). There is no record of the restoration of those
books and papers, and of the powder, which were not returned on
this occasion; but one of these books, the one out of which the minutes
of the action of 10 April 1586 had been cut, is now certainly in the
British Museum (MS. Cotton, Appendix XLVI); the powder of
transmutation is mentioned as being again in Kelly's hands in an action
of 18 April 1587 (Casaubon, p. 11 [of a second pagination, beginning
at the end of the book]).

[Note: Josten's selection is from Ashmole MS 1790 art I, fols. 1-10.]

The conjuring of angels was not Dee's sole concern during these years, for
on 14 May 1586 we find him writing on political and religious matters to
Sir Francis Walsingham, then Principal Secretary of State. Frances Yates has
perceived in Dee's travels a 'Continental Mission', which opinion will be
considered in due course. The fact remains that Dee aroused the hostility
of the newly appointed Papal Nuncio, Philip, Bishop of Piacenza, who accused
him of necromancy in a document submitted to the Emperor who, on 29
May 1586, issued a decree expelling Dee, his family, and Kelley from the
Kingdom of Bohemia and other dominions of the Emperor. However, on
8 August 1586, the Emperor granted to Prince William of Orsini-Rosenberg-
Crumlau licence for them to return to Bohemia under that Prince's protection.
The Prince became a friend, disciple, and patron, and Dee and Kelley went
to his castle at Trebona, where they remained for about two years.

Our next extract begins just after the entry of a Madimi very different
from the creature we first met in the second extract. Now she is all naked
'and sheweth her shame also'. The passage ends with the notorious proposal
that Dee and Kelley share their wives in common.

Saturday April 18, 1587: Trebona

EK: She kneeleth and holdeth up her hands.

Mad: The Laws of God and of His Son Christ, established by the
testimony of his Disciples and Congregation and by the force and power
of his holy Spirits, are not in any particular vocation abrogated but
rather confirmed.[1]

For oftentimes it falleth out that God being offended at the
wickedness of any man, or of some man private, sendeth down his
Spirit of Death, infecting and tempting another man's mind; so that
he becometh void of Reason and riseth up against him, whom God

is offended with, and striketh him so that he dyeth. This, before man, is accounted sin; before God it shall be imputed unto him for righteousness. Even so, whatsoever the Spirit of God teacheth us from Him, though it appear sin before man, is righteousness before him.

Therefore assure yourselves that whatsoever is seen and heard amongst you, is from above, and is a sign and testimony even this day before you; for I that touched thy Son[2] might also have taken away his breath.

But O, you are of little understanding: but behold I teach you.

That unto those that are accounted righteous (through the goodwill of God) sin is justly punished, but not as unto the wicked. For whatsoever you have done unto other men, even the self-same shall light upon you, but happy is he that receiveth not justice through the terror of malediction, but through the grace and mercy of God.

The Apostle Paul[3] abounded in carnal lust: he was also offensive unto his brethren so that he despaired and was ready to have left his vocation, until the Lord did say unto him: *My mercy and grace sufficeth thee.*

Believe me, that we are from above.[4]

Which considered, consider also, that as you cannot comprehend the heavens, so likewise can you not comprehend the wisdom of God which saith: *I will be merciful unto whom I list and unto whom I will not I have none in store: foolish is he that asketh why.*[5]

And behold I say unto you: *Stumble not against God.* Who He is that made you? Who is He that hath given you power to look up towards heaven? You are fools and of little understanding. This day saith God unto you:

Behold you are become free. Do that which most pleaseth you. For behold, your own reason riseth up against my wisdom.

Not content you are to be heirs, but you would be Lords, yea Gods, yea the Judgers of the heavens. Wherefore do even as you list, but if you forsake the way taught you from above, behold, evil shall

---

[1]   A privilege granted doth not abrogate a Law but doth notify the force of the law in itself otherwise.

[2]   Arthur was smitten in a swound and EK saw one in a long, white garment make as though he would smite him. He was very sick for the time.

[3]   Saul of Tarsus lecherous.

[4]   Good Angels.

[5]   The wisdom of God, of us not comprehensible.

enter into your senses and abomination shall dwell before your eyes, as a recompense unto such as you have done wrong unto. And your wives and children shall be carried away before your face.

The Almighty God of heaven and earth be my comfort, as I desire comfort in his service; and give me wisdom as I desire it for His honour and glory, Amen.

EK: I see a white pillar;[6] and upon the pillar I see four heads.

She tieth the pillar round about with a list.

The four heads are like on two heads and on two Wolves' heads. Now there cometh a thing like a white Crown of Crystal and standeth upon all our four heads. The heads seem to be enclosed by the necks within the pillar.

Now she taketh the pillar and goeth up with it.

Now she bringeth an half Moon down and written in it as followeth:

Injustum nihil quod justum est Deo.

Now she goeth round about upon a thing like a Carpet; she goeth now beyond where is an Orchard; she cutteth branches of two trees, and she seemeth to insert them, or graft them into another.

Now she goeth into a black place behind the wood and bringeth a thing with her in a chain; an ugly thing like a Devil.

Mad: Behold, seest thou this: wherewithall thou thoughtest to overthrow and most infect, thou art utterly overthrown and shalt never return again.

EK: Now he leapeth and the ground openeth and he sinketh in: and there seemeth a stink of brimstone to come to my nose from the pit.

Now the grafts are all grown in the tree, as if they were all of one tree.

Now she cometh out of that orchard. Now she goeth round about the orchard and leaveth a darkness like a cloud round about the orchard.

Mad: Visible to God but invisible to man.

EK: Now she cometh again upon her Carpet.

Mad: Behold, if you resist not God but shut out Satan (through unity amongst you)[7] then it is said unto you: Assemble yourselves together every seventh day that your eyes may be opened and that you may

[6]   The Crystalline pillar.
[7]   Unity.

understand by him that shall teach you, what the secrets of the holy books (delivered you) are: that you may become full of understanding and in knowledge above common men.[8]

And in your works go forward and detract no time, that you may also have fruit.[9]

Unto William[10] I will be merciful for ever, according to my promise. But I will buy him no Kingdom, after the manner of man, with money. But what I have determined unto him, shall happen unto him: and he shall become mighty in me.

And this Powder which thou has brought here, is appointed for a time by God and cannot be used until then without offence. Happy is he that heareth my words this day: and happy is he that understandeth them.

But if you deny the Wisdom of the Highest and account us, His Messengers, creatures of darkness . . . this day you are made free.

And look that you lay up all things that is spoken of from above: and whatsoever hath been taught you (as well the books as instruments).

You shall shortly have to do again with the cruelty of the Emperor and the accursed Bishop.

Whereunto, if you go forward with God, you shall be taught to answer. If you leave off, as soon as you hear of it be going into Germany, lest you perish before then.

I have no more to say unto you, but my swiftness is from above.

EK: Now she maketh herself ready, &c.
Mad: If my friendship like you not, I beseech God send you as goodwill as I (in power) bear towards you.

I have not one word more given me to speak.
EK: Now she is gone.

JD: I was glad that an offer was made of being every seventh day to be taught the secrets of the books already delivered unto us: thinking that it was easy for us to perform that unity which was required amongst us four, understanding all after the Christian and godly sense. But E.K. who had yesterday seen and heard another meaning of this unity required, utterly abhorred to have any dealing with them farther, and

---

[8] An offer of every 7th day to be taught the secrets of the books received. The holy books delivered.

[9] Our works to go on.

[10] The Lord Rosenberg.

did intend to accept at their hands the liberty of leaving off to deal with them any more: which, his understanding, as it was strange and unpleasant unto me, so I earnestly requested to be resolved therein in manner as followeth.

At the same time and in the same place, this ensued.

NOTE

JD: Upon Mr Kelly his great doubt bred unto me of Madimi, her words yesterday, spoken to him, *that we two had our two wives in such sort as we might use them in common,* it was agreed by us, to move the question, whether the sense were of carnal use (contrary to the law of the Commandment) or of Spiritual love and charitable care and unity of minds, for advancing the service of God.

EK: Upon a Scroll, like the edge of a Carpet, is written:

De utroq; loquor

JD: The one is expressly against the Commandment of God: neither can I by any means consent to like of that Doctrine. And for my help in that verity, I do call down the power of Almighty God, the Creator of heaven and earth, and all the good Angels (His faithful Ministers) to assist me in the defence of my faithful obedience to the law of the Gospel and of His Church.

> *Assist me, O Christ.*
> *Assist me, O Jesu.*
> *Assist me, O Holy Spirit.*

Did the wife-swapping actually take place despite the understandable reluctance of Jane Dee? The answer lies in *The Complete Enochian Dictionary* (London, 1978) by Donald C. Laycock, who states: 'In the original manuscripts there is a section, heavily erased and barely legible, recording the seance of 23 May 1587 – the morning after the wife-swapping. The spirits ask Kelley: "Was thy brother's wife [Jane Dee] obedient and humble unto thee?" – and Kelley replies: "She was." Dee returns the same answer concerning Joanna Kelley.'

Stephen Skinner[8] points out that one curious feature which followed this strange incident was an improvement in the literary quality of the angelic communications. The reader may judge from the following example, in which the present editor finds passages of striking beauty.

Saturday May 23. Mane Circa 9½

Preces ad Deum fundebantur, &c. And then we requested that the act of obedience performed (according to our faith conceived of our vocation from the Almighty and Eternal God of heaven and earth) might be accepted: and that henceforward we might be instructed in the understanding and practice of wisdom, both such as already we have received some introductions mystical and also of all other what the Almighty God shall deem meet for us to know and execute for his honour and glory &c.

JD: EK took pen and ink and wrote the request here adjoined; and he read it to me, and he requested me to read it to the Divine Majesty; and so I did, and hereupon we waited both to the first my prayer and to this Petition, the Divine answer.

Omnipotens sempiterne, vere & vive Deus mittos lucem tuam & veritatem tuam ut ipsa nos ducant & perducant admontem, sanctum Syon, ex hac valle miserie & ad Celestem tuam Jerusalem. Amen.

EK: From the beginning of this our coming, there appeared a purple Circle as big as a star in the circumference of the holy Stone, which yesterday was brought again: and that it should so be, Madimi had forewarned EK when she shewed it unto him, when also she gave the prints of the letters of the backside of the bootom of the gold frame of it.

There appeareth here a great man all in bright harness sitting upon a white horse: he hath a spear all fiery in his left hand, he not putteth into his right hand: he hath a long sword by his side: he hath also a target hanging on his back, it seemeth to be of steel. It hangeth from his neck by a blue lace, it cometh up behind him as high as the top of his head. The horse is milk-white, all studded with white; a very comely horse it is. The man is in complete harness, the top of his helmet hath a sharp form.

Upon his Target are many Cherubins, as it were painted in Circles: there is one in the middle. About it as a Circle with six in it, and then a Circle with eight, and then a great Circle with ten in it and in the greatest are twenty; and about the Circle of twenty are seven parts, at each of which points is a Cherubin. Their faces be like burning gold, their wings be more brighter and as it were . . . their wings coming over their heads do not touch together. His horse is also harnessed before and behind. The horse's legs behind are harnessed as with boots marvellously contrived, for defence, as it were of his hind legs.

He is ridden away, he seemeth to ride through a great field.

Here is now come Madimi.
She is gone into the field, that way which he rode.
Here is another, like a woman all in green.
Here cometh another woman. All her attire is like beaten gold.
She hath on her forehead a Cross crystal, her neck and breast are bare
unto under her dugs. She hath a girdle of beaten gold slackly buckled
under her with a pendant of gold down to the ground.

THE ANGEL: I am the Daughter of Fortitude and ravished every hour
from my youth. For behold, I am Understanding, and Science dwelleth
in me; and the heavens oppress me, they covet and desire me with
infinite appetite: few or none that are earthly have embraced me, for
I am shadowed with the Circle of the Stone and covered with the
morning Clouds. My feet are swifter than the winds and my hands
are sweeter than the morning dew. My garments are from the beginning
and my dwelling place is in myself. The Lion knoweth not where
I walk, neither do the beasts of the field understand me. I am deflowered
and yet a virgin. I sanctify and am not sanctified. Happy is he that
embraceth me: for in the night season I am sweet and in the day,
full of pleasure. My company is a harmony of many Cymbals and my
lips sweeter than health itself. I am a harlot for such as ravish me and
a virgin with such as know me not. For lo, I am loved of many, and
I am a lover to many; and as many as come unto me as they should
do have entertainment. Purge your streets, O ye sons of men, and
wash your houses clean; make yourselves holy and put on righteousness.
Cast out your old strumpets and burn their clothes; abstain from the
company of other women that are defiled, that are sluttish, and not
so handsome and beautiful as I, and then will I come and dwell amongst
you: and behold, I will bring forth children unto you, and they shall
be the Sons of Comfort. I will open my garments and stand naked
before you, that your love may be more enflamed toward me.

As yet, I walk in the Clouds; as yet, I am carried with the Winds,
and cannot descend unto you for the multitude of your abominations
and the filthy loathesomeness of your dwelling places. *Behold these four:* [1]
*who is he that shall say; They have sinned? or unto whom shall they make
account? Not unto you, O you sons of men, nor unto your children; for unto
the Lord belongeth the judgement of His servants.*

---

[1]   i.e. JD, EK and their wives.

Now, therefore, let the earth give forth her fruit unto you, and let the mountains forsake their barrenness where your footsteps shall remain. Happy is he that saluteth you and cursed is he that holdeth up his hands against you. *And power shall be given unto you from henceforth to resist your enemies: and the Lord shall always hear you in the time of your troubles.*[2] And I am sent unto you to play the harlot with you and am to enrich you with the spoils of other men. Prepare for me, for I come shortly. Provide your Chambers for me, that they may be sweet and cleanly; for I will make a dwelling-place amongst you: and I will be common with the father and the son, yea and with all them that truly favoureth you: for my youth is in her flowers and my strength is not to be extinguished with man. Strong am I above and below, therefore provide for me: for behold, I now salute you, and let peace be amongst you; for I am *the Daughter of Comfort*. Disclose not my secrets unto women, neither let them understand how sweet I am, for all things belongeth not to every one. I come unto you again.

EK: She is gone along that green field also.

JD: I read it over to our great comfort.

JD: We most humbly and heartily thank thee, O God Almighty, the only fountain of Wisdom, Power and all goodness. Help us now and ever to be faithful and fruitful servants to thee, for thy honour and glory, Amen.

EK: The field appeareth a very level ground, covered with pretty grass even to the brinks of the . . . It is bright if the Sun light, but I see not the Sun, but the clear sky over it.

JD: *Pausa semihora unius.*

EK: Now cometh the horseman and rideth by into the field and so doth Madimi. Now cometh the third, and so goeth away into the field.

Now cometh she that was left here: she standeth still: she hath a book in her hand covered (as it were) with moss three inches at the head, and four inches long and a finger thick. It hath no clasps; *it is plain.*

Pausa

Angel: The fourth hour after dinner, repair hither again: and whatsoever you shall read out of this book, receive it kneeling upon your knees; and see that you suffer *no Creature female to enter within this place.* Neither

---

[2]   A blessing for obedience according to faith.

shall the things that be opened unto you *be revealed unto your wives or unto any Creature as yet:* for I will lie with you awhile and you shall perceive that I am sweet and full of comfort, and that *the Lord is at hand,* and that He will *shortly visit the earth,* and all his whole Provinces.

EK: She turneth herself into a thousand shapes of all Creatures and now she is come to her own form again.
She hangeth the Book in the air.
Give God thanks, and so depart.
JD: All laud, thanks, honour and glory be to our God, our King and Saviour, now and ever. *Amen.*

1587 Saturday the same day

After dinner, about four hours or somewhat less, we resorted to the place. A voice to EK: *Kneel towards the East;* so he kneeled at the table of Covenant with his face toward the East; and I at my table opposite to him.

JD: In the Name of God the Father, God the Son and God the Holy Ghost, Amen.
   EK: The Book remaineth hanging in the air.
   A voice: Kelly, I know it is troublesome for thee to kneel: sit.
   JD: So EK rose from kneeling and did sit.
   EK: Now she is here that last advertised us. She taketh the book and divideth it into two parts: and it seemeth to be two books: the half cover adjoining to one, and the other half cover belonging to the other; the sides with the covers are towards me.

The Angel: Wisdom is a piercing beam, which is the centre of the spiritual being of the Holy Spirit, touching from all parts from whence the Divinity sendeth it out: and is proper to the soul or unto substances that have beginning, but no ending; so that whatsoever shall have end, can never attain unto that which is called Wisdom. Neither can things that are subject to the second death, receive any such influence, because they are already noted and marked with the seal of destruction. Happy is he whom God hath made a vessel of salvation; for unto him belongeth joy and a crown of reward. *Adam* (your forefather and first parent) in respect of his creation, that is to say, in respect of his imaginative composition, received no strength but by the Holy Ghost; for the soul of man is free from all passions and affections, until it enter into the body unto the which it is limited: so that, being neither good nor

bad (but apt unto both) he is left by Divine providence and permission
joined together to the end of one, or the other: But wheresoever wisdom
dwelleth, it dwelleth not with the soul, as any property thereof, but
according to the goodwill of God, whose mercy concurreth on every
side into Him, and taketh up a mansion therein, to utter out and manifest
His great goodness. And even as the heavens are glorified continually
with the Spirit of God, so is the soul of man glorified that receiveth
sanctifications thereby; for no man is illuminated that is not sanctified:
neither is there any man perfectly sanctified that is not illuminated.
I speak this (my brethren) for that you shall understand, That no man
did or can ever attain to wisdom (that perfect wisdom which I speak
of) without he become a Center in his soul unto the mercies and goodwill
of God comprehending him and dwelling in him; therefore lift up
your eyes and see, call your wits together and mark my words: to
teach you or expound unto you the mysteries of the Books[3] that you
have already received, is not in my power, but in the goodwill of God,
after whose Image I am. *Which goodwill of God is the descending of His
Holy Spirit abundantly upon you and into you,* opening all your senses
and making you perfect men: for *Adam* understood by that grace and
his eyes were opened so that he saw and knew all things that were
to his understanding. So have all those more and less, that have been
counted wife, received the gifts of the Holy Ghost, which setteth the
soul on man *so on fire that he pierceth into all things and judgeth mightily.
The Apostles which knew even the thoughts of men, understood all things,
because the Holy Spirit made a dwelling place in them: even so shall it happen
unto you. For you are the chosen of these last days and such as shall be* full
of the blessings of God, and his Spirit shall rest with you abundantly.
Mark therefore what I have to say unto you.

A hundred days are limited unto you during the which time, you
shall every seventh present yourselves in this place and you shall laud
and praise God. And behold I will be present amongst you.

And before these days pass, when power is given me so to do,
I will enter out of this Stone unto you[4] and you shall eat up these
two books, both the one and the other: and wisdom shall be divided
between you, sufficient to each man.

Then shall your eyes be opened to see and understand *all such things*

---

[3] The conceiving the exposition of our former books.
[4] Personal apparition.

as have been written unto you and taught you from above.[5] But beware
ye take heed that you dwell within yourselves, *and keep the secrets of
God,*[6] *until the time come that you shall be bid SPEAK.* For then shall
the Spirit of God be mighty upon you, so that it shall be said of you,
LO were not these the Sorcerors and such as were accounted Vagabonds? Other
some shall say: *Behold let us take heed and let us humble ourselves before
them. For the Lord of Hosts is with them.*

    And you shall have power in the Heavens and in the lower bodies. And
it shall be taught to you *at all times inwardly,* even what *belongeth to
the hearts of men.* Then shalt thou EK have a new coat put on thee
and it shall *be all of one colour.*[7] Then shalt thou JD *also have power
to open that book, which God has committed unto thee;* but use yourselves
as men, yea even then remember such as may receive the mercies and
grace of God. *And let all peace and unity be amongst you.* For even as
the Sun looketh into all things from above, *so shall you into all the creatures
that live upon the earth:* yea, the one of you shall have his . . . lifted
and shall enter into the fourth or fifth heaven[8] for unto him that is
worldly, knowledge be given: and unto him that hath been patient,
shall greater things descend.[9]

    Notwithstanding both sufficiently satisfied, in the mean season,
The seventh day hence, shall thou bring in such things, *as the Lord hath
given thee:*[10] *And in this place they shall be disposed according to the knowledge
that is given me.* And herein thou hast pleased the Lord, for that *thou
hast dealt straight and according to brotherly meaning.*

    Now cometh the time that the Whore shall be called before the
Highest, and the tenth month hence shall the Turk and the Moscovite
make a perpetual league together, and in the thirteenth month, shall
Poland be assaulted, with the Tartarians, and shall be spoiled; yea, even
unto the very ribs, so that in the sixteenth month *they shall fall all
together from Christ.* And the hand of God shall run in vengeance,
vengeance, *even through this Kingdom,* and through Germany and into

---

[5]   The understanding of such things as have before time been delivered us
    mystically.

[6]   Silence until Power to be given us.

[7]   *Viae actionem anno 1583, mensis die,* of his divers spotted coat.

[8]   Note and remember Entrance in the 4th and 5th heaven.

[9]   O Lord, I thank thee that thou hast accepted my patience.

[10]   *Maii 30:* the books of *Dunstan* and the powder.

Italy; and in the 23 month, Rome shall be destroyed, so that one stone shall not be left standing upon another, *and vengeance shall be on all the earth,* and fear upon all people, for the Lord is gone out against them. *They eat and drink and say: let us be merry.* Woe be unto them, for they know not the time of their visitation. For lo, Justice shall visit them and tread them under foot: and even this Kingdom shall endure for a while; that is to say, this wicked triumph. And behold in the North shall rise that Monster, and shall pass forth with many Miracles,[11] but you seeing all these things shall be at quietness *until such times as it shall be said unto them, Revenge.* Happy is he that is not partaker of the love of such as shall be vexed these latter days.

EK: She is gone.

JD: I read these over to EK. To his great comfort.

On 10 May 1588 Dee writes: 'EK did open the great secret to me, God be thanked.' J. W. Hamilton-Jones was convinced that this refers to Dee's search for the process of alchemical transmutation.[9]

Certainly Kelley was passionately interested in alchemy. His efforts in this direction restored him to the favour of the Emperor Rudolf and according to Charlotte Fell Smith,[10] Dee was reduced to the unenviable position of friendly supplicant to Kelley. After Dee's return to England, Kelley was knighted by Rudolf II for his alchemical efforts, though the Emperor finally imprisoned Kelley for the latter's failure to produce gold. Kelley's attempt to escape during November 1595 terminated in his falling from a turret and he died from his injuries.

It seems appropriate at this point to mention another legend of this bizarre period. The tale gained currency in occult circles for many centuries and is repeated by Aleister Crowley in his *Magick Without Tears:*[11] 'In the British Museum (and I suppose elsewhere) you may see the medal struck to commemorate the victory over the Armada. This is a reproduction, perhaps modified, of the Talisman used by Dee to raise the storm which scattered the enemy fleet.'

What can finally be said about Dee's angel magic? The first possibility, based upon the fact that Dee himself never actually saw anything and endorsed by Frances Yates,[12] is that Kelley was a fraudulent knave and Dee was his unwitting dupe. There are difficulties in accepting so simple and obvious an explanation. For instance, Kelley appears to have been wholly sincere in his pursuit of alchemy, whether or not Dee was with him, and as Dr Head writes:

[11] Anti-Christ.

There is no question that Kelley had a broad streak of opportunism: but also from the beginning we find him openly doubting the nature of his spiritual contacts, protesting that their nature is diabolical and not angelic. He tells Dee that they are deluders, that his 'heart standeth against them', that their promises cannot be relied upon. During the sittings he is constantly on the alert to catch the spirits out and embarrass them. On one occasion he convicts them of plagiarising from Cornelious Agrippa. [13]

It is also very doubtful whether a mere trickster could have endured the boredom of taking down the Enochian alphabet for fourteen months.

The second possibility is that Kelley was made enough to see visions and Dee was fool enough to take them seriously. 'Though Kelley has usually been pictured as a blatant charlatan – even worse than Dee – it is very difficult to believe that he saw nothing at all,' writes Peter French and adds: 'Dee stresses constantly, "*The key of Prayer* openeth all things"'; and one is struck by the immense piety and the almost continuous state of prayer that inform the spiritual conferences. [14] There is also the small point that Dee used scryers other than Kelley; and the much larger point that an ordered, complex, and coherent system rarely issues from a disordered mind.

Nor was Dee quite the credulous idiot some have seen him as being. As Head puts it: 'As for Dee himself, it is simply not the case that he meekly accepted everything he was told. Most of the time he is a model of caution. He notes every question and every answer; and if a discrepancy appears, he demands that it be explained before going on. He is all humility when praying to God – but in the matter of revelation he is more than ready to "try the spirits whether they are of God".' [15] The final nail in the coffin of this argument has been thudded home by Donald C. Laycock: 'Only a man clinically insane – and Dee was certainly not that – would have filled many hundreds of pages, covering more than two decades in all, with a private fantasy that became revealed to the world only by accident.' [16]

The third possibility is that Kelley was able to activate the (still) little known powers of the unconscious. Perhaps, in the end, it all comes down to Hamlet's words to Horatio: 'There are more things in heaven and earth . . . than are dreamt of in your philosophy.' [17]

9

THE PRIVATE DIARY
[1587]

OUR next selection is from *The Private Diary*, ed. James O. Halliwell, and is included to show other concerns of Dee during this period. A number of interesting points are raised by certain entries.

Oct 28: Was the furnace for alchemical operations?

Nov 21: Is it worth noting that Dee was still in touch with his old friend Edmund Dyer.

Nov 24: An example of the hostility Dee could arouse.

Dec 12: What exactly was the 'spirit' that 'burnt all that was on the table where it stands'? Is Dee referring to the overturning of a spirit lamp? But the 'spirit of wine' is not inflammable unless Dee is referring to brandy. Even so, the ensuing conflagration, which consumed no less than four books, is astonishing. Nor can one see how a fire could 'cast' a book 'on the bed hard by from the table'.

Jan 13: This makes it clear that Dee and Kelley are no longer sharing the same residence and that Kelley's situation has improved sufficiently for him to send his brother as a messenger to beckon Dee to his home 'to pass the tyme with him at play'.

Oct 12th, Mr E. K. toward Prage on horsbak. Oct. 13th, mane paulo ante ortum solis observavi radio astronomico inter . . . . . et . . . . . gradus 2 minuta prima 22, et erat . . . . . sub Tauro in eadem linea perpendiculari ante oculum demissa super horizonta altitudo erat vix quatuor graduum. Oct. 15th, hyred Nicolas. Oct. 20th, I toke up the furniture for the action. Oct. 26th, Mr Edward Kelly cam to Trebona from Prage. Oct. 28th and 29th, John Carp did begyn to make furnaces over the gate, and he used of my rownd bricks, and for the yern pot was contented now to use the lesser bricks, 60 to make a furnace. Oct. 31st, Ed. Hilton cam to Trebona in the morning. Nov. 8th, E. K. terribilis expostulatio, accusatio, &c. hora tertia a meridie. Nov. 17th, John Basset had seven ducketts beforehand for his second quarter's wages, begynning the 1st. Nov. 21st, Saterday at night Mr Francis Garland cam from England to Trebona and browght me a letter from Mr Dyer and my brother Mr Richard. Nov. 24th, at the marriag super Critzin the Grand Captayn disdayned to com thither to supper in the Rad howse of Trebona becawse E. K. and I were there; and sayd farder that we were . . . . . Dec. 1st to 11th, my Lord lay at Trebon and my Lady all this tyme. Dec. 10th, Mr John Carpio went toward Prage to marry the mayden he had trubbled; for the Emperor's Majestie, by my Lord Rosenberg's means, had so ordred the matter. Dec. 12th, afternone somwhat, Mr Ed. Keley his lamp overthrow, the spirit of wyne long spent to nere, and the glas being not stayed with buks abowt it, as it was wont to be; and the same glas so flitting on one side, the spirit was spilled out, and burnt all that was on the table where it stode, lynnen and written bokes,—as the bok of Zacharius with the Alkanor that I translated out of French for som by spirituall could not; Rowlaschy his thrid boke of waters philosophicall; the boke called Angelicum Opus, all in pictures of he work from the beginning to the end; the copy of the man of Badwise Conclusions for the Transmution of metalls; and 40 leaves in 4°, intitled, Extractiones Dunstani, which he himself extracted and noted out of Dunstan his boke, and the very boke of Dunstan was but cast on the bed hard by from the table.

1588. Jan. 1st, abowt nine of the clok afternone, Michel, going chilyshly with a sharp stik of eight ynches long and a little wax candell light on the top of it, did fall uppon the playn bords in Marie's chamber, and the sharp point of the stik entred throwgh the lid of his left ey toward the corner next the nose, and so persed throwgh, insomuch that great abundance of blud cam out under the lid, in the very corner of the sayd eye; the hole on the owtside is not bygger than a pyn's hed; it was anoynted with St John's oyle. The boy slept well. God spede the rest of the cure! The next day after it apperid that the first towch of the stikes point was at the very myddle of the apple of the ey, and so (by God's mercy and favor) glanced to the place where it entred; with the strength of his hed and the fire of his fulness, I may make some shew of it to the prayse of God for his mercies and protection. Jan. 11th, Nicolas was sore hart circa 8½ hora nocte. Jan. 13th, at dynner tyme Mr Edward Kelly sent his brother, Mr Th. K. to me with these words, "My brother sayth that you study so much, and therfor, seeing it is too late to go to day to Cromlaw, he wisheth you to come to pass the tyme with him at play."

[From *The Private Diary,* ed. James O. Halliwell, Camden Society Publications, vol. XIX (London, 1982).]

10

LETTER TO
QUEEN ELIZABETH
[1588]

WHY was Dee so often abroad, and for what purposes? Why did he need the permission of Sir William Cecil to remain in Antwerp in 1563? Why did Leicester and Walsingham direct him to Frankfurt-on-Oder in 1578? One interesting solution has been proposed by Richard Deacon in *John Dee: Scientist, Geographer, Astrologer and Secret Agent to Elizabeth I* (London, 1968). Deacon sees Dee as incorporating among his many roles that of 'a roving James Bond of Tudor times'.

Wayne Shumaker responds acidly to this claim: 'Dee was not the CIA man that Richard Deacon portrays in his worthless fiction . . . Deacon has all Dee's gullibility but none of his learning.'[1] Peter French concurs: 'In a rather sensational way, Deacon portrays Dee as . . . the master of a massive espionage system. He considers the *Spiritual Diaries* a form of enciphering used for spying purposes. Deacon's argument is tenuous at best and his book is riddled with factual inaccuracies.'[2] These scholarly opinions should be borne in mind when reading Deacon's book. Even so, the questions Deacon raises do remain and more research is needed on Dee's interest in cryptography.

Frances Yates sees in Dee's travels 1583-9 a complex 'Continental Mission' which was religious in nature and 'vast, undogmatic, reforming'.[3] Unfortunately, the evidence for this is scanty, though persuasively marshalled by Yates. Again, more scholarship is required.

Our next extract, the *Letter to Queen Elizabeth* (1588), is an attempt to prepare the ground for what Dee hoped would be a triumphant return.

Most gracious Sovereign Lady, the God of heaven and earth Who has mightily and evidently given unto your most excellent Royal Majesty the wonderful, triumphant victory against your mortal enemies, be always thanked, praised and glorified; and the same God Almighty evermore direct and defend your most Royal Highness from all evil and encumbrance: and finish and confirm in your most excellent Majesty Royal the blessings long since both decreed and offered: yea, even into your most gracious Royal bosom and lap. Happy are they that can perceive and so obey the pleasant call of the mighty Lady, *Opportunity*. And therefore finding our duty concurs with a most secret beck of the said Gracious Princess, Lady Opportunity, NOW to embrace and enjoy your most excellent Royal Majesty's high favour and gracious great clemency of CALLING me, Mr Kelley and our families home into your British Earthly Paradise and Monarchy incomparable: (and that, about a year since; by Master Customer Yong, his letters); I and mine (by God, his favour and help and after the most convenient manner we can) will, from henceforth, endeavour ourselves, faithfully, loyally, carefully, warily and diligently, to rid and untangle ourselves from hence. And so, very devoutly . . . at your Sacred Majesty's feet, to offer ourselves, and all wherein we are or may be able to serve God and your most excellent Royal Majesty. The Lord of Hosts be our help and guide therein: and grant unto your most excellent Royal Majesty the Incomparablest Triumphant Reign and Monarchy that ever was, since Man's creation. Amen.

[Trebon, Bohemia, 10 November 1588; Harleian MS 6986, fol. 45.]

PERSONAL DIARIES
[1589–95]

WE ARE now in a position to form some estimation of the character of Dr John Dee.

Certainly he had his failings. In common with most men, he was vain in his achievements and often felt unappreciated. His inordinate ambition drove him to dream of grandiose schemes which aroused the suspicion of practical people. Perhaps his most tedious fault was a lack of humour, for he never wrote a line intended to provoke laughter, and this inhibited his understanding of other points of view: he never could suffer fools gladly. At all times he took himself extremely seriously, which many must have found irritating.

In common with many who dedicate their lives to the arts and sciences, Dee valued money as a tool for the advancement of his work, not as a thing to be valued in itself, and without a thought for the morrow he spent freely on his labours. Thus he was periodically short of money and the problem grew ever more acute in his later years. Nor did he possess much talent for repairing his fallen fortunes. Whether it was selling his church livings for a song or declining offers of patronage which could have solved all his financial problems, Dee usually made the wrong decision.

On the other hand, he was probably not consciously responsible for a characteristic some might censure: a morbid suspicion that enemies were plotting against him. His imprisonment under Queen Mary, which could have led to his being burned at the stake, made a lasting impression upon him; nor should one forget that in 1600 Dee's brilliant contemporary, Giordano Bruno, would meet his end in the flames of the Inquisition. As Dee grew older, he became increasingly possessed by the conviction that enemies were ruining his credit behind his back in a furtive orgy of lies.

Why did Dee style himself 'Doctor'? It has been suggested that the degree was conferred at Louvain, but Charlotte Fell Smith denies this possibility.[1] W. I. Trattner thought the title honorary since Dee was 'doctus, or learned'.[2] However, C. H. Josten has presented evidence[3] to support the claim that Dee was a doctor of medicine, receiving the degree from the University of Prague around 1585/6. Peter French[4] agrees that Dee was knowledgeable about medicine but points out that John Foxe refers to Dee as doctor long before the date of the degree awarded by Prague University. [5]

There is also the matter of Dee's ancestry. He firmly believed himself to be descended from Roderick the Great, ancient Prince of all Wales, and a distant relative of Queen Elizabeth I. One would like to know if this was the case.

Whether these claims were justified or not, Dee's vices are trifling compared to his virtues. He was the polymathic intellectual giant of his era, committed to purity of thought in a courageous endeavour to penetrate and explore all spheres of knowledge, known and unknown, visible and invisible. High-minded to an excessive degree, perhaps, he was nevertheless broad-minded too, and as a sincere Christian numbered Protestants and Catholics among his friends, hoping with all his heart for a future wise and tolerant unity within Christendom.

Vanity and pride were among his failings, yet one is conscious also of a genuine sense of honour and dignity. He appears to have been that rare individual, a truly honest man. Throughout his work he is consistently well intentioned, praying earnestly that his plans might bring about greatness for his country, with liberty, justice, peace and prosperity for one and all.

He impressed great contemporaries. Judging from the number and diversity of his social engagements and from the duration of his personal relationships, he must have been a good and sociable friend. 'A very handsome man', according to Aubrey, [6] he liked women enough to marry three times, and apart from the wife-swapping incident dictated (as he believed) by the angels themselves, he seems to have been sexually monogamous. Again according to Aubrey, he was a good neighbour: 'He was a great Peace-maker; if any of the neighbours fell out, he would never lett them alone till he had made them friends.' [7]

Yet his good qualities were to be of little avail during the ensuing years when he sought for patronage with growing desperation, as his *Diaries* 1589-95 make clear. Certain entries require our attention.

16 Dec 1589: It is obvious that at this point, Dee remains optimistic and enjoys encouragement from the Queen, who gives her approval to his alchemical studies.

2 March 1591: Dee is very short of money. This growing concern forces him into the unenviable position of humble petitioner, subject to continual delays and disappointments.

22 Nov 1592: Dee reads his *Compendious Rehearsal* to the Queen's Commissioners and we shall be looking at this in the next chapter.

29 June 1594: A frustrated Dee records a bitter disappointment.

18 April 1595: Dee is finally granted the wardenship of Manchester College. Since the appointment is very far from the Court, one can safely assume that Dee has fallen from royal favour.

The editing, which is once again by Hippocrates Junior (pseud),[8] shows the commencement of Dee's decline into poverty and obscurity, from which it has taken four centuries to rescue him. The reasons for this decline will be commented upon further in later chapters: for the present we recall Aubrey's verdict: 'a mighty good man he was'.[9]

1589

Oct. 31st Letters sent to Stade for Gerwein Greven for her Majestie, Mr Yong, and Mr Dyer.

Dec. 19th At Richemond with the Quene's Majestie.

1590

July 14th Mr Gawayn Smyth spake frendely for me to the Quene, and she disclosed her favor toward me.

Nov 20th Her Majestie cam to Richemond. Nov. 27th, the Quene's Majestie, being at Richemont, graciously sent for me. I cam to her at three quarters of the clok afternone, and she sayd she wold send me something to kepe Christmas with. Dec. 1st, Her Majestie commanded Mr John Herbert, Master of Requests, to write to the Commissioners on my behalf. Dec. 2nd, order taken by the Commissioners for my howse and goods. Her Majestie told Mr. Candish that she wold send me an hundred angels to kepe my Christmas withall. Dec. 4th, the Quene's Majestie called for me at my dore, *circa 3½ a meridie,* as she passed by, and I met her at Estshene gate, where she graciously, putting down her mask, did say with mery chere, 'I thank thee, Dee; there was never promise made but it was broken or kept.' I understode her Majestie to mean of the hundred angels she promised to have sent me this day, as she yesternight told Mr. Richard Candish. Dec. 6th . . . A meridie circa 3ª recipi a Regini Domina 50*l.* Dec. 14th, the Quene's Majestie called for me at my dore as she rod by to take ayre, and I met her at Estshene gate. Dec. 16th, Mr. Candish receyved from the Quene's Majestie warrant by word of mowth to assure me to do what I wold in philosophie and alchimie, and none shold chek, controll, or molest me; and she sayd that she wold ere long send me 50*l.* more to make up the hundred pound.

1591

Mar. 2nd   Borrowed 20l. uppon plate, and payd this day 19l. in Mortlak.

Dec. 20th   A jentle answer of the Lord Threasorer that the Quene wold have me have something at this promotion of bishops at hand.

1592

Mar. 6th   The Quene granted my sute to Dr. Awbrey.

Nov. 9th   Her Majestie's grant of my supplication for commissioners to comme to me. The Lord Warwik obteyned it. Nov. 22nd, the commissioners from Her Majestie, Mr. Secretary Wolley and Sir Thomas George, cam to Mortlak to my howse . . . Dec. 1st, a little after none the very vertuous Cowntess of Warwik sent me word very speedily by hir gentleman Mr. Jones from the cowrt at Hampton Cowrt that this day Her Majestie had granted to send me spedily an hundred marks, and that Sir Thomas George had very honorably dealt for me in the cause. Dec. 2nd, Sir George Thomas browght me a hundred marks from Her Majestie.

Feb. 15th   Her Majestie gratiously accepted of my few lynes of thankfulnes delivered unto her by the Cowntess of Warwik, *hora secunda a meridie,* at Hampton Court, two or three dayes before the remove to Somerset Howse.

1594

April 1st   Capitayn Hendor made acquayntance with me, and shewed me a part of his pollicy against the Spanishe King his intended mischief agaynst her Majestie and this realme. May 3rd, betewne 6 and 7 after none the Quene sent for me to her in the privy garden at Grenwich, when I delivered in writing the hevenly admonition, and Her Majestie tok it thankfully. Onely the Lady Warwyk and Sir Robert Cecill his Lady wer in the garden with Her Majestie. May 18th, Her Majestie sent me agayn the copy of the letter of G. K. with thanks by the Lady Warwick. May 21st, Sir John Wolley moved my sute to Her Majestie. She graunted after a sort, but referred all to the Lord of Canterbury. May 25th, Dr Awbrey moved my sute to Her Majestie, and answere as before. May 29th, with the Archbishop before the Quene cam to her house. June 3rd, I, my wife, and seven children, before the Quene at Thisellworth. My wife kissed her hand. I exhibited my request for the Archbishop to com to my cottage. June 6th, supped with the Lord Archbishop. Invited him to my cottage.

June 29th After I had hard the Archbishop his answers and discourses, and that after he had byn the last Sonday at Tybald's with the Quene and Lord Threasorer, I take myself confounded for all suing or hoping for anything that was. And so adiew to the court and courting tyll God direct me otherwise!

Dec. 7th Jane my wife delivered her supplication to the Quene's Majestie, as she passed out of the privy garden at Somerset Howse to go to diner to the Savoy to Syr Thomas Henedge. The Lord Admirall toke it of the Quene. Her Majestie toke the bill agayn and kept [it] uppon her cushen; and on the 8th day, by the chief motion of the Lord Admirall, and somwhat of the Lord Buckhurst, the Quene's wish was to the Lord Archibishop presently that I shold have Dr Day his place in Powles.

1595

Jan. 3rd The Wardenship of Manchester spoken of by the Lord Archbishop of Canberbury. Feb. 5th, my bill of Manchester offered to the Quene afore dynner by Sir John Welly to signe, but she deferred it.

Apl. 18th My bill for Manchester Wardenship signed by the Quene, Mr Herbert offring it her.

July 31st The Cowntess of Warwik did this evening thank her Majestie in my name, and for me, for her gift of the Wardenship of Manchester. She toke it gratiously; and was sorry that it was so far from hens, but that some better thing neer hand shall be fownd for me; and, if opportunitie of tyme wold serve, her Majestie wold speak with me herself. I had a bill made by Mr Wood, one of the clerks of the signet, for the frutes given me by her Majestie.

Oct. 9th I dyned with Syr Walter Rawlegh at Durham Howse.

THE COMPENDIOUS
REHEARSAL
[1597]

ONE of the many qualities which distinguished the Renaissance was the quest for harmony of thought, word, and deed. To take a mundane example, in Elizabethan England a man of substance aspired to explore unknown parts of the Earth, fight the Queen's enemies, make love to ladies, appreciate a scientific instrument, and turn a neat sonnet. The same quest prevailed in the sphere of the intellect, and the intelligent gloried in the liberated powers of the independent human mind. This ideal had its enemies and as the sixteenth century came to a close, these grew in number, influence, and power. One result was the eventual ruination of John Dee.

Dee no longer enjoyed a prestigious place at Court and the influence of his friends and supporters was on the wane. Leicester's abortive expedition to the Netherlands in 1586 led to the death of Sir Philip Sydney and the former's disgrace, closely followed by his own death in 1588. Edmund Spenser, poet of Dee's philosophy and introduced at Court by Raleigh in 1592, was sent away to Ireland, returning to London in 1599 to die in poverty and neglect. By the time Sir Walter Raleigh dined with Dee in 1595, he himself was out of favour.

Dee's *Compendious Rehearsal,* first read to the Queen's Commissioners in 1592 and printed in 1597, is an attempt to justify his life, enumerate his achievements, stress his loyalty, patriotism and piety, compel sympathy for his sufferings and evoke royal patronage. Four extracts have been chosen to exemplify these points.

[CHAPTER V]

SOME MY DUTIFULL SERVICES DONE UNTO HER MAJESTIE IN THE SPACE OF THIRTY-FOUR YEARES AND MORE

1. Before her Majesties coming to the crowne, I did shew my dutifull good will in some travailes for her Majesties behalfe, to the comfort of her Majesties favourers then, and some of her principall servantes, at Woodstock, and at Milton by Oxford, with Sir Thomas Bendger

(then Auditor unto her Majestie), and at London; as Mr Richard Strange and Mr. John Asheley, now Master of her Majesties Jewell house, might have testified, and as I could have brought to their remembrance.

Upon suspicion of which my service then, and upon the false information given in by one George Ferrys and Prideaux, that I endeavored by enchantmentes to destroy Quene Mary, I was prisoner at Hampton Court, even in the weeke next before the same Whitsontide, that her Majestie was there prisoner also. I remained long prisoner, and all dores of my lodgings in London sealed up; and with other circumstances of griefe, loss, and discredit for a while endured under the keeping of diverse overseers: as first in Court under Sir John Bourne, Secretary: while by writing I answered first four articles, and thereupon eighteen other, administered unto me by the right honourable the Privy Councell. Then from thence I was sent on Whitesun-even with the guard by water to London to the Lord Broke, Justice of the Common Pleas; from thence at length to the Star Chamber: where I was discharged of the suspicion of treason, and was sent to the examining and custody of Byshop Bonner for religious matters. Where also I was prisoner long, and bedfellow with Barthlet Grene, who was burnt: and at length upon the King and Queenes clemency and justice, I was (A. 1555, Augusti 19) enlarged by the Councell's letters; being notwithstanding first bound in recognizance for ready appearance and the good abearing for about some four moneths after; which letter of the Councells is in print here to be scene: as the forepart of this narration may be seen in the records of the Councell Chamber of that yeare, month, and day, if they be extant.

2. Before her Majesties coronacion I wrote at large, and delivered it for her Majesties use by commandement of the Lord Robert, after Earle of Leicester, what in my judgment the ancient astrologers would determine of the election day of such a tyme, as was appointed for her Majestie to be crowned in. Which writing, if it be extant and to be had, wilbe a testimony of my dutifull and carefull endeavour performed in that, which in her Majesties name was enjoyned me: A. 1558.

3. Her Majestie tooke pleasure to heare my opinion of the comet appearing A. 1577: whereas the judgment of some had unduly bred great feare and doubt in many of the Court; being men of no small account. This was at Windsore, where her Majestie most graciously,

for three* divers daies, did use me; and, among other pointes, her most excellent Majestie promised unto me great security against any of her kingedome, that would, by reason of any my rare studies and philosophicall exercises, unduly seeke my overthrow. Whereupon I againe to her Majestie made a very faithfull and inviolable promise of great importance. The first part whereof, God is my witnes, I have truly and sincerely performed; though it be not yet evident, how truly, or of what incredible value: The second part by God his great mereyes and helpes may in due tyme be performed, if my plat for the meanes be not misused or defaced.

4. My carefull and faithfull endeavours was with great speede required (as by divers messages sent unto me one after another in one morning) to prevent the mischiefe, which divers of her Majesties Privy Councell suspected to be intended against her Majesties person, by meanes of a certaine image of wax, with a great pin stuck into it about the brest of it, found in Lincolnes Inn fields, &c., wherein I did satisfie her Majesties desire, and the Lords of the honourable Privy Councell within few houres, in godly and artificiall manner: as the honourable Mr Secretary Willson, whome, at the least, I required, to have by me a witnes of the proceedings: which his Honor before me declared to her Majestic, then sitting without the Privy Garden by the landing place at Richmond: the honorable Earle of Leicester being also by.

5. My dutifull service was done, in the diligent conference, which, by her Majesties commandment, I had with Mr Dr Bayly, her Majesties Physitian, about her Majesties grievous pangs and paines by reason of toothake and the rheume, &c. A. 1578, October.

6. My very painefull and dangerous winter journey, about a thousand five hundred myles by sea and land, was undertaken and performed to consult with the learned physitians and philosophers beyond the seas for her Majesties health-recovering and preserving; having by the right honourable Earle of Leicester, and Mr Secretary Walsingham but one hundred dayes allowed unto me to goe and come againe in, A. 1578. My pas-port here may somewhat give evidence, and the journall litle book of every dayes journey or abode for those hundred dayes account may suffice.

7. My great, faithfull, and carefull attendance about the Lady

___

* One of three daies at Windsor Mrs Skydarior, now Lady Skydarior, hath some remembrance.

Marquess of Northampton (A. 1564) both beyond the seas, on the seas, and here in England, was performed with her Majesties good will and well liking of. Whereupon her Majestie was the more willing, at the suite of the said Lady Marquiss, to give unto me, for some recompence, the deanery of Glocester; but I was disappointed, as I have before specified, of the enjoying of it.

8. My faithfull diligence and earnest labour, with some cost, was bestowed, by her Majesties commandment, to set down in writing, with hydrographicall and geographicall description, what I then had to shew or say, as concerning her Majesties title royall to any forreine countries. Whereof, the two parchment great rolls full writtin, of about xii whole velome-skinns, are good witnes here before you. For coppy whereof I have refused an hundred poundes in money offred by some subjectes of this kingdome: but it was not meete for me to take it.

9. My dutifull labout, commanded by her Majestie, upon the Gregorian publishing of a Reformation of the vulgar Julian yeare, may here appeare unto you in these two written bookes, having ben read and examined by learned mathematitians (therto assigned by the honourable Lords of the Councell) and by their skylls also warranted; and by the Lords of the Councell and by the Barons of the Exchequer well liked off, for the manner of execution of it without any publique cumber or damage, &c. A. 1582.

10. I sent very dutifully, humbly, and faithfully out of Bohemia (A. 1585) letters unto her sacred Majestie, requesting an expert, discreet, and trusty man to be sent unto me in Bohemia, to heare and see, what God had sent unto me and my friendes there at that tyme; at which tyme, and till which tyme, I was chief governour of our philosophicall proceedings; and by both our concentes, there was somewhat prepared and determined upon to have ben sent unto her Majestie, if the required messenger had been sent by her Majestie unto us. But not long after (so soone as it was perceived, that my faithfull letters were not regarded therein) by lithe and lithe I became hindered and crossed to performe my dutifull and chiefe desire; and that, by the fyne and most subtill devises and plotts laid, first by the Bohemians, and somewhat by Italians, and lastly by some of my owne countrymen. God best knoweth how I was very ungodly dealt withall, when I meant all truth, sincerity, fidelity, and piety towardes God, and my Quene and country.

And so to conclude this chapter: if in any other pointes, besides the forerehearsed, I have done my dutifull service any way to her Majesties well liking and gracious accepting, I am greatly bound to thanke Allmighty God, and during my lyfe to frame the best of my little skill to doe my bounden duty to her most excellent Majestie. [*marginal note:* 'Her sacred Majestie best knoweth my sincere, zealous, constant, and dutifull fidelity towardes her.']

[from CHAPTER X]

The particular true accounts of all these gifts, loanes, and debts upon skore, talley, or booke, are here before your Honours; which I beseech you to looke over, and to consider of them especially, how the usury devoureth me, and how the skore, talley, and booke debts doe dayly put me to shame in many places and with many men; some of them having been required of me, or at my house, very often tymes within these four moneths. What can I answer to such without shame or great griefe of minde? But where the fault lyeth, God Almighty he knoweth, and he chiefly can redress it.

Of which former totall summe of £833 your Honour seeth here, how much hath been paid in extraordinary debts and expenses, as one hundred pounds to enjoy my house in quiet (as before tyme) by the order: £40 my debt at Breme: £20 already for usury, &c., and soe in all £267, which being deducted from the £833, leaveth the ordinary charges of house keeping for these three yeares last, to have been £566, and that with great parsimony used. And hereunto must be added the value of many gifts and helps for our house keeping sent to me by good friends; as vessells of wine, whole brawnes, sheepe, wheat, pepper, nutmegg, ginger, sugar, &c., and other things for the apparell of me, my wife, and our children. The value of all which to be rated in mony, may well be judged to have been above £50, which added to the former £566, yeeldeth the total summe of the charges of house keeping to have been £616. Whereof undoubtedly the £600 hath been spent in meate, drinke, and fuell, and the other 16 odd pounds may be allowed for the apparell stuff, bestowed on us by guift.

Thus your honours most briefly do perceive, how mercifully our God hath been unto us, and bountifull in his provision hitherto, by preparing and enclining the charitable hearts and ready hands of some true Christians, her Majesties loving subjects, to cloath the naked, and to feede the hungry, &c. Of which charitable points and some more I and myne have tasted in these three last yeares: wherein the Almighty hath tryed our faith and trust in him and in his word. And even he hath now opened the eyes and eares of her most gracious Majestie, and inclined her most mercifull heart to see, heare, and grant the most humble supplication of her true, faithfull, dutifull, and obedient servant: to whose most bountifull and most gracious speedy succour and reliefe, both the almighty God for his part, and I for myne, doe committ the whole cause at this present, upon your honours convenient report made therein unto her most excellent Majestie.

[CHAPTER XI]

MY LAMENTABLE AND FINALL COMPLAINT, AND MOST FERVENT PETITION

Seeing therefore by God his most secret providence and purpose, with his helping hand, I have (very patiently and with good hope) used and enjoyed the five former honest and lawfull means of provision makeing to preserve my selfe, my wife, our children, and family from hunger-sterving and nakednes the space of these whole and just three* yeares last past; and that notwithstanding have in the meane space very often made declaration to our superiours of the great distress and incredible want, unto which I was unduely brought; and partly by reason, that her most gracious Majesties favour towards me and her express commandment royall for my sufficient mainteynance and reliefe-enjoying, hath not hitherto byn so regarded, as any fruit, or one penny yearely revenue is thereby assured unto me yet:

And seeing noe one way of all the five former manners of remedies (which I have been forced to use for these three yeares, for the most needefull mainteynance of me, my wife, children, and family); seeing no one, no one (I say the third tyme) of those five waies doth now remain any longer to be enjoyed by any seemely order to our knowledge, and that my onely house is left to be sold [*marginal note:* 'And even now am constrained to mortgage my house for £100 onely, to pay presently my debts, grown on me forcibly within these 4 last yeares onely.'] outright, and that for halfe the money it cost me, wherewith to pay some of my debts and not all: what, I beseech your Honors, may I doe or shall I doe lawfully henceforward; whereby I may prevent, that I and myne shall not be registered in chronicles or annalls to the

---

* and now 4 yeares and 5 moneths last past.

posterity of true students for a warning not to follow my steps; and thereby to be soe unjustly, unchristianly, and unnaturally so long forced and driven to such very disgracefull shifts and full of indignities (as they may full well be termed, if my former declaration be duly considered); yea, at length to be left remediless of such inconvenient shifts also, and to be brought even to the very next instant of stepping out of dores (my house being sold) I and myne, with botles and wallets furnished to become wanderers as homish vagabonds; or, as banished men to forsake the kingdome?

But if a number, accounted students in this kingdome, who can spend yearely by ecclesiasticall livings four hundred pounds in value or revenue, as for reward of their well doing, or for maintenance of their studies, should be constrain'd to leese those their four hundred pound revenues to me; unless they could with six daies warning ballance downe the pith of this dayes my declaration, with a better of theirs in all respects, undoubtedly then should I not be long destitute of sufficient ability and mainteynance for me and myne.

Therefore seeing the blinded lady, Fortune, doth not governe in this commonwealth, but *justitia* and *prudentia,* and that in better order, then in Tullie's *Republica* or bookes of Offices they are laied forth to be followed and performed, most reverently and earnestly (yea, in manner with bloody teares of heart), I and my wife, our seaven children, and our servants (seaventeene of us in all), do this day make our petition unto your Honors, that upon all godly, charitable, and just respects had of all that, which this day you have scene, heard, and perceived, you will make such report unto her most excellent Majestie (with humble request for speedy reliefe), [*marginal note:* 'Mark I pray you.'] that we be not constrained to do or suffer otherwise, then becometh Christians, and true, faithfull, and obedient subjects to doe or suffer; and all for want of due mainteynance.

[CHAPTER XII]

THE RESOLUTION FOR GENERALL, VERY EASY, AND SPEEDY REMEDY, IN THIS RARE AND LAMENTABLE CASE

Undoubtedly, her most excellent Majesties gracious grant by word of mouth, yea four times within these three last yeares already pronounced in my behalfe of the Mastership of St Crosses, or the Wardenship of Winchester, or Provostship of Eaton, or Mastership of Sherborn, or such like, being speedily performed and assured unto me, and of me enjoyed, may be a sufficient remedy against all inconveniences, otherwise most likely to ensue: the extreame pinch of all manner of want for meat, drinke, fewell, cloth, &c., incredibly tormenting me and myne at this present, after three yeares continuall my very hard getting and making of provision for our most needeful mainteynance, even to the uttermost and last meanes used therein: alwaies notwithstanding with great good hope (from moneth to moneth) that, in respect of her most excellent Majesties very great favour towards me; and in respect of her most gracious and expresse commandement divers tymes by word of mouth and letter declared therein; I should ere this have been otherwise and more charitably regarded: and so some sufficient and certaine reliefe and mainteynance should have been bestowed on me ere this.

[From *Autobiographical Tracts of Dr John Dee,* ed. James Crossley, Chetham Society Publications, vol. XXIV (Manchester, 1851).]

13

LETTER TO THE ARCHBISHOP OF CANTERBURY
[1599]

IT HAS been stressed that central to the Renaissance was the will to dignify Man's place in the scheme of things and restore his awareness of his divine nature, bringing him ever closer to God. This was the moving spirit behind its occult philosophy. Yet, with the fascinating exception of Giordano Bruno, occult philosophers regarded themselves as sincere Christians and insisted that their teachings were entirely compatible with Christianity. Even so, many Christians did not believe them. [1]

The occult philosophy could co-exist peacefully with the humanist movement for Catholic reform as exemplified by Erasmus, More, and Colet in the early years of the sixteenth century. It could also be reconciled with the moderate Protestantism which made faith and the dictates of conscience supreme. But it was incompatible with the intolerance, bigotry, and religious fanaticism which contaminated the century's closing years.

Catholic and Protestant men of goodwill sought for an acceptable compromise which would restore Christian unity. There was hope that something of this sort might be achieved at the Council of Trent, but the result was a victory for intolerance. In England, a reaction against the excesses of the Protestant Northumberland and the Catholic Mary led to the Elizabethan Settlement, an endeavour to create a Broad Church wedded to the State and to which all Christians could comfortably subscribe; but this was soon enough threatened both by Counter-Reformation Catholics and the growing number of strict Puritans.

All this was awkward for John Dee. One cannot doubt either the sincerity of his Christian beliefs or his desire for universal religious harmony. Frances Yates has argued that the latter desire motivated Dee's travels in Europe 1583-9. [2] Certainly, though Dee was a Protestant, he numbered Catholics among his friends, and though religious zeal abounds in his writings, there is no evidence of religious bigotry.

All visions of universal religious harmony perished in the fires of the Counter-Reformation. The Index of prohibited books denied faithful Catholics access to information which might cause them to question the Church's teaching on any matter. The revived Inquisition tortured and slaughtered heretics, silencing Galileo and consigning Dee's great contemporary Bruno to the flames in Rome in 1600. Men like Dee, who earnestly desired peace, must have been appalled by the French wars of Religion and the cruelty of Catholic Spain's attempts to crush the Dutch Revolt, though in the succeeding century even these horrors would be eclipsed by the atrocities both sides committed in the Thirty Years War.

Dee's *Letter Containing a most briefe Discourse Apologeticall* addressed to the Archbishop of Canterbury (1599) displays his awareness of a growing peril. There follows an extract from the *Compendious Rehearsal* in which Dee lists his works. In the ensuing 'Fervent Protestation' and 'Epilogue', the reader may detect a note of panic.

TO THE MOST REVEREND FATHER IN GOD, THE LORD ARCHBISHOP OF CANTERBURY, PRIMATE AND METROPOLITANE OF ALL ENGLAND, ONE OF HER MAJESTY'S MOST HONORABLE PRIVY COUNCILLERS: MY SINGULAR GOOD LORD.

Most humbly and heartily I crave your Grace's pardon, if I offend any thing, or send or present to your Grace's hand so simple a Discourse as this is: although in the opinion of some sage and discreet friends, it is thought not to be impertinent, to my most needful suits presently in hand (before her most Excellent Majesty Royal, your Lordship's good Grace, and other of the Right Honourable Lords of Her Majesty's Privy Council) to make some part of my former studies and studious exercises (within and for these 46 years continued) known to your Grace and other Right Honourable good Lords of her Majesty's Privy Council.

And secondly, afterwards the same to be permitted to come to public view. Not so much to stop the mouths and at length stay the impudent attempts of the rash and malicious devisers and contrivers of most untrue, foolish and wicked reports and fables concerning my aforesaid studious exercises, with my great (yea, incredible) pains, travails, cares and costs in the search and learning of true Philosophy–as to certify and satisfy the godly and impartial Christian hearer or reader thereof. That by his own judgement (upon his due consideration and examination of this (no little parcel) of the particulars of my aforesaid studies and philosophical exercises) he will or may be sufficiently informed and persuaded that I have wonderfully laboured to find, follow, use and haunt the true straight and most narrow path, leading all true, devout, zealous, faithful and constant Christian students . . .

All thanks are most due, therefore, unto the Almighty, seeing it pleased Him (even from my youth, by His divine favour, grace and help) to insinuate into my heart an insatiable zeal and desire to know His truth. And in Him and by Him incessantly to seek and listen after the same, by the true Philosophical method and harmony proceeding and ascending (as it were) *gradatim,* from things visible to consider things invisible: from things bodily, to conceive of things spiritual: from things transitory and momentary, to meditate on things permanent: by things mortal (visible and invisible) to have some intimation of immortality.

And, to conclude most briefly by the most marvellous frame of the whole World, philosophically viewed and circumspectly weighed, numbered and measured (according to the talent and gift of GOD, from above alotted, for His Divine Purpose effecting) most faithfully to love, honour, and glorify always the *Framer* and *Creator* thereof. In Whose workmanship, His infinite goodness, inscrutable wisdom and Almighty power may (by innumerable means) be manifested and demonstrated.

The truth, of which my zealous, careful and constant intent and endeavour is here specified, may (I hope) easily appear by the whole, full and due survey and consideration of all the Books, Treatises and Discourses, whose titles only are at this time here annexed and expressed. They are set down in the sixth chapter of another little rhapsodical treatise, entitled *The Compendious Rehearsal,* written over two years ago, for Her Majesty's two honourable Commissioners, which Her Most Excellent Majesty most graciously sent to my poor cottage in Mortlake, to understand the matter and the causes in full; through which I was so extremely urged to procure at Her Majesty's hands such Honourable Surveyors and witnesses to be assigned, for the due proof of the contents of my most humble and pitiful supplication, exhibited to Her Most Excellent Majesty, at Hampton Court on 9th November 1592. Thus therefore (as follows) is the said sixth chapter:

My labours and pains, bestowed at divers times, to pleasure my native country, by the writing of sundry books and treatises: some in Latin, some in English, and some of them written at Her Majesty's commandment.

Of which books and treatises, some are printed and some unprinted. The printed books and treatises are as follows:

1- *Propaedeumata Aphoristica.* (1558)

2– *Monas Hieroglyphica*. (1564)

3– *Epistola Ad Eximium Ducis Urbini Mathematicum* (Fredericum Commandinum) *praefixa libello Machometi Bagdedini, De Superficierum Divisionibus*. (1570)

4– *The British Monarchy* (otherwise called *The Petty Navy Royal*). (1576)

5– My *Mathematical Preface*, annexed to *Euclid* . . . wherein are many Arts wholly invented by me (by name, definition, propriety and use), more than either the Greek or Roman Mathematicians have left for our knowledge. (1570)

6– My divers and many *Annotations* and *Mathematical Inventions,* added in sundry places of the aforesaid *Euclid* after the tenth book. (1570)

7– *Epistola prefixa Ephemeridibus* Johannis Felde *Angli, cui rationem declaraveram Ephemerides conferibendi*. (1557)

8– *Paralaticae Commentationis, Praxeosq, Nucleus quidam*. (1573)

The unprinted books and treatises are these: (some perfectly finished and some yet unfinished).

9– The first great volume of famous and rich discoveries: wherein (also) is the History of King Solomon, every three years his Ophirian voyage: the originals of Presbyter John: and of the first great Cham and his successors for many years following: the description of divers wonderful isles in the Northern, Scythian, Tartarian and the other most Northern Seas and near under the North Pole. By Record, written over 1200 years ago: with divers other rarities. (1576)

10– The British Complement, of the Perfect Art of Navigation. A great volume in which are contained: our Queen Elizabeth, her Arithmetical Tables Gubernautic: for Navigation by the Paradoxall Compass (invented by me in 1557) and Navigation by great Circles: and for longitudes and latitudes, and the variation of the Compass, which finds the true direction most easily and speedily, yea (if need be), in one minute of time, and sometimes without sight of Sun, Moon or Star: with many other new and needed inventions Gubernautic. (1576)

11– Her Majesty's Royal Title to many foreign Countries, Kingdoms and Provinces, by good testimony and sufficient proof recorded: and in 12 vellum skins of parchment fair written, for Her Majesty's use and at Her Majesty's commandment. (1578)

12– *De Imperatoris Nomine, Authoritate & Potentia*. Dedicated to Her Majesty. (1579)

13- Prologomena & Dictata Parisiensia, in *Euclidis* Elementorum Geometricorum; librum primum & secundum. (1550)

14- De Usu Globi Coelestis: ad Regem Edoardum Sextum. (1550)

15- *The Art of Logic,* in English. (1547)

16- *The Thirteen Sophistical Fallacians,* with their discoveries, written in English meter. (1548)

17- Mercurius Coelestis: libri 24, written at Louvain. (1549)

18- De Nubium, Solis, Lunae, acreliquorum Planetarum, immoipsius stelliseri Coali. (1551)

19- Aphorismi Astrologici 300. (1553)

20- The true cause and account (not vulgar) of Floods and Ebbs: written at the request of the Right Honourable Lady, Lady Jane, Duchess of Northumberland. (1553)

21- The Philosophical and Poetical Original Occasions of the Configurations, and names of the heavenly Asterisines, written at the request of the same Duchess. (1553)

22- The Astronomical & Logical Rules and Canons to calculate the Ephemerides by; and other necessary accounts of heavenly motions: written at the request and for the use of that excellent Mechanician, Master Richard Chancellor, at his last voyage into Moschovia. (1553)

23- De Acribologia Mathematica. (1555)

24- Inventum Mechanicum, Paradoxum, De nova ratione delineandi Circumferentiam Circularem . . . (1556)

25- De Speculis Comburentibus . . . (1557)

26- De Perspectiva illa, qua peritissimi utuntur Pictores. (1557)

27- Speculum unitatis: five Apologia pro Fratre *Rogerio Bacon:* in qua docetur nihil illum per Daemoniorun fecisse auxilia, sed philosophum fuisse maximum; naturaliterque & modis homini Christiano licitis, maximas fecisse res, quas indoctun solet vulgus, in Daemoniorum referre facinora. (1557)

28- De Annuli Astronimici multiplici usu; *lib 2.* (1557)

29- Trochillica Inventa, *lib 2.* (1558)

30- Περὶ αγαβιβασμῶν δεολογιχῶν *lib 3.* (1558)

31- De tertia & praecipua Perspective parte, quae de Radiorum fractione tractat, *lib 3* (1559)

32- De Itinere subterraneo, *lib 2.* (1560)

33- De Triangulorum rectilineorum Areis, *lib 3.* (1560)

34- Cabalae Hebraicae compendiosa tabella. (1562)

35- Reipublicae Britanicae Synopsis: in English. (1565)

36- De Trigono Circinoque Analogico, Opusculum Mathematicum & Mechanicum, *lib 4.* (1565)

37- De stella admiranda, in Cassiopeae Asterismo, coelitus demissa ad orbem usque Veneris: Iterumque in Coeli penetralia perpendiculariter retracta, post decimum sextum suae apparitionis mensem. (1573)

38- Hipparchus Redivivus, Tractatulus. (1573)

39- De unico Mago, & triplici *Herode,* eoque Antichristiano. (1570)

40- Ten sundry and very rare Heraldical Blazonings of one Crest or Cognizance, lawfully confirmed to certain ancient Arms, *lib 1.* (1574)

41- Atlantidis (vulgariter, Indiae, Occidentalis nominatae) emendatior descriptio Hydrographica, quam ulla alia adhuc evulgata. (1580)

42- De modo Evangelii Jesu Christi publicandi, stabiliendique, inter Infideles Atlanticos. (1581)

43- Navigationis ad Carthayum per Septentrionalia Scythiae & Tartariae litora, Delineatio Hydrographica: *Arthuro Pit* & *Carlo Jackmano* Anglis, versus illas partes Navigaturis, in manus tradita. (1580)

44- Hemisphaerii Borealis Geographica, atque Hydrographica descriptio. (1583)

45- The Originals and chief points of our ancient British Histories discoursed upon and examined. (1583)

46- An advice and discourse about the Reformation of the vulgar Julian year, written by Her Majesty's commandment, and the Lords of the Privy Council. (1582)

47- Certain considerations and cofferings together of these three sentences, anciently accounted as Oracles: *Nosce teipsum: Homo Homini Deus: Homo Homini Lupus.* (1592)

48- De hominis Corpore, Spiritu & Anima: sive Microcosmicum totius Philosophiae Naturalis Compendium, *lib 1.* (1591)

With many other books, pamphlets, discourses, inventions and conclusions, in divers Arts and matters, whose names need not be noted in this Abstract. The most part of all which, specified here, lie before your Honours upon the Table, on the left hand side. But by other books and writings of another sort (if it so please God and if He will grant me life, health and due maintenance thereto, for some ten or twelve years to come) I may hereafter, without a doubt, make this sentence plainly come true: *Plura latent, quam patent.*

Thus far (my good Lord) have I set down this catalogue, out of the aforesaid sixth chapter of the book whose title is:

49- *The Compendious Rehearsal of John Dee, his dutiful declaration and proof of the course and race of his studious life, for the space of half an hundred years, now (by God's favour and help) fully spent, &c.*

To which *Compendious Rehearsal,* an *Appendix* now belongs concerning these last two years, in which I have had many just occasions to confess that *Homo Homini Deus* and *Homo Homini Lupus* was and is an Argument worthy of being deciphered and discussed at length; and may one day (by God's help) be published in some very strange manner. And besides all the aforesaid books and treatises I have written, I have just cause of late to write and publish a treatise with the title: *De Horizonte Aeternitatis,* to make evident that one Andreas Libavius, in a book of his printed last year, has insufficiently considered a phrase of my *Monas Hyeroglyphica,* and has a hostile attitude towards it due to his lack of skill in the matter, not understanding my apt application thereof in one of the most important passages of the book. And this book of mine, by God's help and favour, shall be dedicated to her most Excellent Royal Majesty. And this treatise contains three books:

1– The first, entitled: *De Horizonte: liber Mathematicus & Physicus.*
2– The second: *De Aeternitate: liber Theologicus, Metaphysicus & Mathematicus.*
3– The third, *De Horizonte Aeternitatis: liber Theologicus Mathematicus & Hierotechnicus.*

(It may now be here also remembered, that almost three years after the writing of this Letter, I did somewhat satisfy the request of an Honourable Friend at Court by speedily penning some matter concerning her Majesty's Sea Sovereignty, under the title: 51– Thalaattocratia Brytannica.)

Truly, I have great cause to praise and thank God for your Grace's very charitable using of me: both in sundry other matters and also in your favourable yielding to, yea, & notifying the due means for the performance of her Sacred Majesty's most gracious and bountiful disposition, resolution and very royal beginning, to restore and give to me (her ancient, faithful servant) some due maintenance to spend the rest of my days in some quiet and comfort, with the ability to retain some speedy, fair and Orthographical writers about me who are skilled in Latin and Greek (at the very least): also for my own books and works to be copied fairly and correctly (such, I mean, as either her most Excellent Majesty may choose and command to be finished or published, or such of them as your Grace shall think fit or worthy for my labour to be bestowed upon): and also for the speedy, fair and true copying out of other ancient authors' good and rare works in Greek or Latin, which by GOD'S Providence have been preserved from the spoilation of my Library and of all my moveable goods in 1583.

In this Library, there were about 4000 books, of which 700 were anciently written by hand: some in Greek, some in Latin, some in Hebrew and some in other languages (as is made clear by the Catalogue). But the great losses and damages of various kinds which I have sustained, do not so much grieve my heart as the rash, lewd and most untrue fables and reports of me and my Philosophical Studies have done and still do: which usually, after their first hatching and devilish devising, are immediately spread through the Realm, and to some seem true; to others, they are doubtful; and only to the wise, modest, discreet, godly and charitable (and chiefly to those who have some acquaintance with me) they appear and are known to be fables, untruths and utterly false reports and slanders.

Well, this shall be my last charitable giving of warning and fervent protestation to my Countrymen, and all others in this case.

A FERVENT PROTESTATION

Before the Almighty GOD and your Lordship's good Grace, this day, on peril of my soul's damnation (if I lie or take his name in vain herein) I take the same GOD to be my witness: That with all my heart, with all my soul, with all my strength, power and understanding (according to the measure thereof which the Almighty has given me)–for the most part of time, from my youth, I have used and still use good, lawful, honest, Christian, and divinely prescribed means to attain to the knowledge of those truths which are fit and necessary for me to know. And to do His Divine Majesty such service as He has, does and will call me to do during this my life; for His honour and glory to be advanced; and for the benefit and public good of this Kingdom; as much (as by the will and purpose of God) as is within my skill and ability to perform–as a true, faithful and most sincerely dutiful servant to our most gracious and incomparable Queen Elizabeth, and as a very comfortable fellow-member of the body politic, governed under the Sceptre Royal of our earthly Supreme Head (Queen Elizabeth) and as a lively, sympathetic and true fellow-member of that holy and mystical body, Catholically extended and played (wheresover) on the Earth.

In the view, knowledge, direction, protection, illumination, and consolation of the Almighty, most blessed, most holy, most glorious, co-majestical, coeternal and coessential Trinity: the Head of that Body being only our Redeemer, Christ Jesus, perfect God and perfect man, Whose return in glory we faithfully await and daily do very earnestly cry unto Him, to hasten His Second Coming for His Elect's sake. Iniquity so abounds on this earth and prevails, and true faith with Charity and evangelical simplicity have but cold, slender and uncertain entertainment among the worldly-wise men of this world.

Therefore (in conclusion) I beseech the Almighty God most abundantly to increase and confirm your Grace's heavenly wisdom and endow you with all the rest of His heavenly gifts for the relieving, refreshing and comforting, both bodily and spiritually, His little flock of the faithful yet militant here on earth.

Amen.

AN EPILOGUE

Good my Lord, I beseech your Grace to allow me my plain and comfortable Epilogue for this matter at this time.

1–Seeing my studious exercises and civil conversation may be abundantly testified, to my good credit, in most parts of all Christendom, and that by all degrees of Nobility, by all degrees of the Learned and by very many others of godly and Christian disposition, for the space of 46 years (as appears in the Records lately viewed by two honourable witnesses, by commission from her Majesty–

2–And feeling that for the past 36 years, I have been her most Excellent Majesty's very true, faithful and dutiful servant, from whose Royal mouth I have never received one word of reproach, but nothing other than favour and grace; in whose princely countenance I never perceived a frown directed at me, or discontented regard; but at all times favourable and gracious, to the great joy and comfort of my true, faithful and loyal heart–

3–And seeing the works of my hands and words of my mouth (noted before in the Schedule of my writings) which bear lively witness of the thoughts of my heart and inclination of my mind generally (as all wise men do know and Christ Himself does avouch)–

It might seem unnecessary, carefully (though most briefly and speedily) to have warned or confounded the scornful, the malicious, the proud and the rash in their untrue reports, opinions, and fables about my studies and philosophical exercises. But it is of more importance that the godly, the honest, the modest, the discreet, grave and charitable Christians (English or other), lovers of Justice, Truth and good Learning, may hereby receive a certain comfort in themselves (to perceive that *Veritas tandem praevalebit*) and be sufficiently armed with sound truth to defend me against such of my adversaries who will subsequently begin afresh or hold on obstinately to their former errors, vain imaginings, false reports and most ungodly slanders of me & my studies.

Therefore (to make this cause forever, before God and man, out of all doubt): Seeing your Lordship's good Grace is, as it were, our High Priest and Chief Ecclesiastical Minister (under our most dread Sovereign Lady Queen Elizabeth) to whose censure and judgement I submit all my studies and exercises; yea, all my books past and present and hereafter to be written by me (with my own skill, judgement or opinion), I do, at this present time, most humbly, sincerely and unfainedly, and in the name of Almighty God (yea, for his honour and glory) request and beseech your Grace . . . that wherever I have used, or do or shall use pen, speech or conversation otherwise than as it appears to a faithful, careful, sincere and humble servant of Jesus Christ, that your Grace would vouchsafe to clear me . . .

[From *The Autobiographical Tracts of Dr John Dee,* ed. by J. Crossley (1851); transliteration: GS.]

14

PERSONAL DIARIES
[1600–1]

GLOOM, despondency, and cultural pessimism prevailed in the closing years of Elizabeth's reign, finding later expression in the Jacobean revenge drama of Webster, Tourneur, Middleton, and Shakespeare. These feelings must have prevailed also in the breast of the neglected John Dee, enduring semi-banishment in Manchester. .

One hopes that there were consolations for him. He was apparently consulted on matters of witchcraft and demonic possession,[1] though, as we shall see, expertise in these perplexed questions would soon become sorely disadvantageous. Perhaps his leisure hours were spent immersed in the pursuit of antiquarianism: this aspect of Dee has been most ably studied by Peter French.[2]

Two points stand out from the *Diaries* 1600–1 as selected here: the slander and quarrels which blighted Dee's declining years (e.g. 18 July and 11 September 1600); and the penury which forced him to borrow small sums to ensure the survival of his family and himself (e.g. 20 December 1600 and 19 January 1601).

The extract is from *The Private Diary,* edited by Halliwell.

1600. June 10th, set out from London. Jun. 18th, I, my wife, Arthur Rowland, Mistres Marie Nicols, and Mr Richard Arnold cam to Manchester.

July 3rd, the Commission set uppon in the Chapter Howse. July 7th, this morning, as I lay in my bed, it cam into my fantasy to write a boke, 'De diffentiis quibusdam corporum et spirituum.' July 8th, I writ to the Lord Bishop of Chester by Mr. Withenstalls. July 10th, Mr Nicols and Barthilmew Hickman cam. July 14th, Francys Nicols and Barthilmew Hikman went homeward. July 17th, I willed the fellows to com to me by nine the next day. July 18th, it is to be noted of the great pacifications unexpected of man which happened this Friday; for in the forenone (betwene nine and ten) where the fellows were greatly in doubt of my heavy displeasure, by reason of their manifold misusing of themselves against me, I did with all lenity interteyn them, and shewed the most part of the things that I had browght to pass at London for the college good, and told Mr Carter (going away) that I must speak with him alone. Robert Leigh and Charles Legh were by. Secondly, the great sute betwene Redishmer and me was stayed and by Mr Richard Holland his wisdom. Thirdly, the organs uppon condition was admitted. And fourthly, Mr Williamson's resignation granted for a preacher to be gotten from Cambridge. July 19th, I lent Randall Kemp my second part of Hollinshed's Great Chronicle for ij. or iij. wekes. To Newton he restored it. July 31st, we held our audit, I and the fellows for the two yeres last past in my absence, Olyver Carter, Thomas Williamson, and Robert Birch, Charles Legh the elder being receyver. I red and gave unto Mistres Mary Nicolls her prayer.

Aug. 5th, I visited the grammar schole, and fownd great imperfection in all and every of the scholers to my great grief. Aug. 6th, I had a dream after midnight of my working of the philosopher's stone with other. My dreame was after midnight toward day. Aug. 10th, Eucharistam suscepimus, ego, uxor, filia Katharina, et Maria Nicolls. Aug. 30th, a great tempest of mighty wynde S.W. from 2 tyll 6, with rayne.

Sept. 11th, Mr Holland of Denby, Mr Gerard of Stopford, Mr Langley, commissioners from the bishop of Chester, authorized by the bishop of Chester, did call me before them in the church abowt thre of the clok after none, and did deliver to me certayn petitions put up by the fellows against me to answer before the 18th of this month. I answered them all eodem tempore, and yet they gave me leave to write at leiser. Sept. 16th, Mr Harmer and Mr Davis, gentlemen of Flyntshire, within four or five myle of Hurden Castell, did viset me. Sept. 29th, I burned before Mr Nicols, his brother, and Mr Wortley, all Bartholomew Hikman his untrue actions. Sept. 30th, after the departing of Mr Francis Nicolls, his dowghter Mistres Mary, his brother Mr William, Mr Wortley, at my returne from Deansgate, to the ende whereof I browght them on fote, Mr Roger Kooke offred and promised his faithfull and diligent care and help, to the best of his skill and powre, in the processes chymicall, and that he will rather do so then to be with any in England; which his promise the Lord blesse and confirm! He told me that Mr Anthony considered him very liberally and frendely, but he told him that he had promised me. Then he liked in him the fidelity of regarding such his promise.

Oct. 13th, be it remembered that Sir Georg Both cam to Manchester to viset Mr Humfrey Damport, cownsaylor of Gray's Inne, and so cam to the colledg to me; and after a few words of discowrse, we agreed as concerning two or three tenements in Durham Massy in his occupying. That he and I with the fellows wold stand to the arbitrement of the sayd Mr Damport, after his next return hither from London. John Radclyf, Mr Damport's man, was with him here, and Mr Dumbell, but they hard not our agrement; we were in my dyning-room. Oct. 22nd, receyved a kinde letter from the Lord Bishop of Chester in the behalfe of Thomas Billings for a curatship. Nov. 1st, Mr Roger Coke did begyn to destill. Nov. 4th, the commission and jury did finde the titles of Nuthurst due to Manchester against Mr James Ashton of Chaterdon. Nov. 7th, Oliver Carter his . . . . . before Mr Birth, Richard Legh and Charles Legh, in the colledg howse. Dec. 2nd, colledg awdit. Allowed my due of £7 yerely for my howse-rent tyll Michelmas last. Arthur Dee a graunt of the chapter clerkship from Owen Hodges, to be had yf £6 wer payd to him for his patent. Dec. 20th, borowed of Mr Edmund Chetam the scholemaster £10 for one yere uppon plate, two bowles, two cupps with handles, all silver, waying all 32 oz. Item, two potts with cover and handells, double gilt within and without, waying 16 oz.

1601. Jan. 19th, borrowed of Adam Holland of Newton £5 till Hilary day, uppon a silver salt dubble gilt with a cover, waying 14 oz. Feb. 2nd, Roger Cook his supposed plat laying to my discredit was by Arthur my sone fownd by chaunce in a box of his papers in his own handwriting circa meridiem, and after none abowt 1½ browght to my knowledg face to face. O Deus, libera nos a malo! All was mistaken, and we reconcyled godly. Feb. 10th to 15th, reconciliation betwene us, and I did declare to my wife, Katharine my dowghter, Arthur and Rowland, how things wer thus taken. Feb. 18th, Jane cam to my servyce from Cletheraw. Feb. 25th, R. K[oke] pactum sacrum hora octava mane. March 2nd, Mr Roger Coke went toward London. March 19th, I receyved the long letters from Bartholomew Hickman hora secunda a meridie by a carryer of Oldham. April 6th, Mr Holcroft of Vale Royall his first acquaintance at Manchester by reason of William Herbert his friend. He used me and reported of me very freely and wurshiply.

15

LETTER TO
KING JAMES I
[1604]

DEE'S situation was already deteriorating sufficiently before the accession to
the throne of James I in 1603. In place of the regal Elizabeth, courted by
soldiers and hymned by poets, there now stumbled a slobbering, bandy-legged,
homosexual pedant. James's *Daemonologie* (1587), which expressed his
abhorrence of witches and his horror at those who denied the ever-present
reality of witchcraft, amply justifies the verdict of his contemporary, King
Henry IV of France: 'the wisest fool in Christendom'.

In all countries infested by the plague of witch-hunting, mere suspicion
fed by rumour was enough to destroy one's credit and, in many cases, one's
life. On the Continent, torture was applied freely to extract confessions and
the records make for sickening reading. It has been estimated[1] that from
the fifteenth to the seventeenth century, no less than 6,000,000 'witches'
were judicially murdered.

The desire to persecute is the hallmark of mental mediocrity and sexual
inadequacy: yet the coronation of James I must have made Dee's enemies
rejoice. Dee was not a witch, but the slandering of his name as a conjurer
of devils left him open to that charge and subject to the death penalty if
convicted. Hence his petition *To the King's Most Excellent Majesty* (1604)[2]
where, in a desperate attempt to clear his reputation, Dee courageously demands
that he be brought to trial and accepts a hideous fate if convicted.

The petition was ignored.

TO THE KING'S MOST EXCELLENT MAJESTY

1604 June 5: Greenwich

In most humble and lamentable manner, your Highness' most distressed
servant beseeches your Royal Majesty . . . to cause your Highness'
said servant to be tried and cleared of that horrible and damnable and
to him, most grievous and defamatory slander, generally and for many
long years raised in this Kingdom and continued against him by report

and print: Namely, that he is or has been a *Conjurer* or *Caller* or *Invocator* of devils. Which most ungodly and false report, so boldly, constantly and impudently asserted, has been uncontrolled and unpunished for so many years (even though your Majesty's said servant has published in print his divers and earnest refutations of it).

Yet some impudent and malicious foreign enemy or English traitor to the flourishing state and honour of this Kingdom has in print (7 Jan 1592) affirmed your Majesty's said suppliant to be the Conjurer belonging to the most honourable Privy Council of your Majesty's most illustrious predecessor, Queen Elizabeth. The said abominable slander is so heinous and disgraceful, for it discredits and brings into hatred the said honourable Lords of your Majesty's Privy Council (for using *any* Conjurer's advice–and your said suppliant to be the man).

It therefore seems (in many respects) to be essential for an Order to be made speedily by your Majesty's wisdom and Supreme Authority . . . to have your Highness' said suppliant tried on these charges, whcih suppliant offers himself willingly to the punishment of Death (yea, either to be stoned to Death: or to be buried alive: or to be burned unmercifully) if by any due, true and just means the said name of *Conjurer* or *Caller* or *Invocator* of devils or damned spirits can be proved to have been or to be duly and justly reported of or attributed to him.

Yes, (good and gracious King), if any *one* of all the great number of the very strange and frivolous fables reported of him and told of him (as to have been of his doing) were True, as they have been told or reasonably caused any wondering among the many-headed multitude of anyone else, he would be guilty. But your Highness' said suppliant (upon his Justification and Clearing of name made by the trial) will conceive great and undoubted hope that your Majesty will, soon after, more willingly have Princely regard for your Highness' said suppliant's redressed reputation, and for his pains and difficulties, which can no longer possibly be endured by him, so long as his utter undoing, little by little, been (most unjustly) planned and executed.

The Almighty and most merciful God always direct your Majesty's royal heart in His ways of justice and mercy, as is to Him most acceptable: and make your Majesty the most blessed and triumphant Monarch that ever this British Empire enjoyed.

Amen.

16

A LATER
SPIRITUAL DIARY

OUR final selection of Dee's writings finds him poor and sad, yet still faithful to his angels.[1] Others might rob him of his fame and fortune but his integrity was not for sale and could not be stolen.

Nevertheless, the value of the communications has declined, along with Dee himself, and we find him questioning Raphael about 'the blood, not coming out of my Fundament, but at a little, as it were a pin hole of the skin . . .' The leader of the Elizabethan Renaissance died in 1608, impoverished, ignored, and subsequently execrated.

Yet Dee was not without significant posthumous influence. The matter has been most thoroughly explored by Frances Yates in *The Rosicrucian Enlightenment*,[2] a brilliant exposition of the roots and results of the abortive 'Rosicrucian Movement' in the early decades of the seventeenth century. In Yates' opinion: 'The German Rosicrucian manifestos reflect the philosophy of John Dee which he had spread abroad in the missionary venture of his second, or continental, period.'[3]

The ideals espoused at the court of the Palatinate by the unfortunate King Frederick and Queen Elizabeth were, Yates argues, those of John Dee. Unfortunately, the Rosicrucian dream lasted but a brief season, for with the coming of the Thirty Years War, the King and Queen were forced to flee into exile and all hope of realizing the Rosicrucian vision in a political state perished in the blood and fire of Catholic reaction. The Protestant response would be no less virulent.

Yates rightly equates the Elizabethan occult philosophy with Rosicrucianism and in England its principal exponent after Dee was Robert Fludd. This tradition persisted into the nineteenth century and continues today.

London   At Mrs Goodman, her house
            March 20, 4.15 pm.

JESUS

Ominipotens sempiterne & une Deus

Mittas lucem tuam & veritatem tuam, ut ipsa me ducat & perducat ad montem sanctum tuum & Tabernacula. Amen.

VOICE: I am blessed *Raphael,* a blessed messenger of the Almighty, I am sent of God, who is blessed for evermore. Amen.

John Dee, I am sent of God for thy comfort first to certify thee thou shalt overcome this thy infirmity, and when thou art strong in body, as God in his goodness will make thee, THEN *thou shalt have all made known unto thee of such things being not come to pass as have been before spoken of,* because that thou shouldst take comfort in God, that thou art not left from the comfort of God's blessed creatures. Now God hath sent me at this time whereby thou shalt be satisfied, THAT when *thy body is able to abide the time of my service from God to be delivered unto thee by me* Raphael: Thy friend *John Pontoys* yet liveth, but his time is likely to be short.

Ask at your will.

J.D.: O God, I am beaten into a great attempt, to make the council privy of my beggary, and to offer the Earl of Salisbury such my duties as I may perfect to his content. How stands this with your good liking?

RAPHAEL: Thou shalt have friends in thy suit and thou shalt have foes, but through God's mercies, thy friends shall overcome thy foes and thou shalt see how God in His goodness will work mightily in his power for thee.

Proceed in thy suit so shortly as thou can find thy health in body able. And for thy health use thy own skill, that God hath and shall guide thee withal to thy good and perfect receiving of thy perfect health.

J.D.: Of the blood, not coming out of my Fundament, but at a little as it were a pin hole of the skin . . .

RAPHAEL: That the, which thou hadst no knowledge to help thy weakness, God in His mercies did send thee therein present help, the which but only for that issue thou couldst not have lived. And for the cure and thy help, the same God will work with thee in thy heart and mind so, that it shall be known unto no man, but by God's merciful goodness delivered unto thee, such ways and means as shall by thy help, and restore thee to health again. This God of His mercy hath sent me to deliver this short message because of thy weakness, *Thou art not strong to endure them,* therefore such is God's goodness to let you to understand that after the tenth day of April, I will then appear again, and thou shalt understand much more what God's will and his pleasure is to be done in God's services, and for your good, and so for this little short message I have declared unto you the will of *Jesus Christ:* And so for this time, *in the Name of the most highest Creator and maker of Heaven and Earth,* I do now return at His will and commandment, and I am ready at all times when He shall command me to appear to thy comfort. His Name be praised evermore. Amen, Amen.

Amen.

[From *A True & Faithful Relation . . .*, ed. Meric Casaubon (London, 1659).]

17

SOME OPINIONS OF DEE
[1600–1700]

I HAVE chosen in conclusion four extracts of seventeenth century writing
in order to demonstrate the diversity of reactions to Dee.

The History of His Life and Times by William Lilly, the celebrated astrologer,
is a racy read. The attentive reader will note Lilly's claim that Dee was 'queen
Elizabeth's intelligencer' and this is conceivably the rock upon which Richard
Deacon built his case.[1] Although the matter requires further research, in
the same passage Lilly states that Dee was 'educated in the university of Oxford'
when he was in fact a Cambridge man and rarely visited Oxford.

In *Brief Lives* by John Aubrey, there is a charming sketch of Dee the man.

In *The Alchemist* (1610), Ben Jonson saw in Dee and his concerns a target
for the exercise of his Puritan scepticism and biting satire. Jonson's attitude
has been analysed deftly by Frances Yates,[2] but a sample should nevertheless
be included, for so many have echoed Jonson's views down the centuries.
The first extract mocks the behaviour of the alchemist; the second pours
scorn on alchemical obscurity; and the third mentions Dee by name.

Finally, there is a selection from *The Tempest,* a play which extols the
role of the Renaissance magus. The case for believing that Prospero was directly
inspired by Dee has been (once again) cogently argued by Frances Yates[3]
and the present editor finds her case persuasive. It is appropriate, therefore,
to leave the last word to William Shakespeare.

(i) from *History Of His Life and Times* by William Lilly (1602–81):

Dr Dee himself was a cambro-briton, educated in the university of
Oxford, there took his degree of doctor; afterwards for many years
in search of the profounder studies, travelled into foreign parts: to be
serious, he was queen Elizabeth's intelligencer, and had a salary for
his maintenance from the secretaries of state. he was a ready-witted
man, quick of apprehension, very learned, and of great judgment in
the Latin and Greek tongues. He was a very great investigator of the
more secret Hermetical learning, a perfect astronomer, a curious

astrologer, a serious geometrician; to speak truth, he was excellent in all kinds of learning.

With all this, he was the most ambitious person living, and most desirous of fame and renown, and was never so well pleased as when he heard himself stiled most excellent.

He was studious in chymistry, and attained to good perfection therein; but his servant, or rather companion, Kelly, out-went him, viz. about the elixir or philosopher's stone; which neither Kelly or Dee attained by their own labour and industry. It was in this manner Kelly obtained it, as I had it related from an ancient minister, who knew the certainty thereof from an old English merchant, resident in Germany, at what time both Kelly and Dee were there.

Dee and Kelly being in the confines of the emperor's dominions, in a city where resided many English merchants, with whom they had much familiarity, there happened an old friar to come to Dr Dee's lodging. Knocking at the door, Dee peeped down stairs. 'Kelly,' says he, 'tell the old man I am not at home.' Kelly did so. The friar said, 'I will take another time to wait on him.' Some few days after, he came again. Dee ordered Kelly, if it were the same person, to deny him again. He did so; at which the friar was very angry. 'Tell thy master I came to speak with him and to do him good, because he is a great scholar and famous; but now tell him, he put forth a book, and dedicated it to the emperor: it is called ''Monas Hierogliphicas.'' He understands it not. I wrote it myself, I came to instruct him therein, and in some other more profound things. Do thou, Kelly, come along with me, I will make thee more famous than thy master Dee.''

Kelly was very apprehensive of what the friar delivered, and thereupon suddenly retired from Dee, and wholly applied unto the friar; and of him either had the elixir ready made, or the perfect method of its preparation and making. The poor friar lived a very short time after: whether he died a natural death, or was otherwise poisoned or made away by Kelly, the merchant, who related this, did not certainly know.

How Kelly died afterwards at Prague, you well know: he was born at Worcester, had been an apothecary. Not above thirty years since he had a sister lived in Worcester, who had some gold made by her brother's projection.

Dr Dee died at Mortlock in Surrey, very poor, enforced many times to sell some book or other to buy his dinner with, as Dr Napier of

Linford, in Buckinghamshire, oft related, who knew him very well.

I have read over his book of 'Conference with Spirits,' and thereby perceive many weaknesses in the manage of that way of mosaical learning: but I conceive; the reason why he had not more plain resolutions, and more to the purpose, was, because Kelly was very vicious, unto whom the angels were not obedient, or willingly did declare the questions propounded; but I could give other reasons, but those are not for paper.

(ii) from *Brief Lives* by John Aubrey (1626–97); written *c.*1667–97):

HEE had a very faire cleare rosie complexion; a long beard as white as milke; he was tall and slender; a very handsome man. His Picture in a wooden cutt is at the end of Billingsley's *Euclid.* He wore a Gowne like an Artist's gowne, with hanging sleeves, and a slitt; a mighty good man he was.

My great Grandfather, Will: Aubrey, and he were Cosins, and intimate acquaintance. Mr Ashmole hath letters between them, under their owne hands, viz. one of Dr W. A. to him (ingeniosely and learnedly written) touching the *Sovraignty of the Sea,* of which J. D. writt a booke which he dedicated to Queen Elizabeth and desired my great grandfather's advice upon it. Dr A.'s countrey-house was at Kew, and J. Dee lived at Mortlack, not a mile distant. I have heard my grandmother say they were often together.

Among the MSS in the Bodlean library of Doctor Gwyn, are severall letters between him and John Dee, of Chymistrey and Magicall Secrets.

Meredith Lloyd sayes that John Dee's printed booke of *Spirits,* is not above the third part of what was writt, which were in Sir Robert Cotton's Library; many whereof were much perished by being buryed, and Sir Robert Cotton bought the field to digge after it. He told me of John Dee, etc., conjuring at a poole in Brecknockshire, and that they found a wedge of Gold; and that they were troubled and indicted as Conjurors at the Assizes; that a mighty storme and tempest was raysed in harvest time, the countrey people had not knowen the like.

Old Goodwife Faldo (a Natif of Mortlak in Surrey) did know Dr Dee, and told me that he did entertain the Polonian Ambassador at his howse in Mortlak, and dyed not long after; and that he shewed the Eclipse with a darke Roome to the same Ambassador. She beleeves that he was eightie years old when he dyed. She sayd, he kept a great many Stilles goeing. That he layd the storme. That the Children dreaded him because he was accounted a Conjurer. He recovered the Basket of Cloathes stollen, when she and his daughter (both Girles) were negligent: she knew this.

He used to distill Egge-shells, and 'twas from hence that Ben: Johnson had his hint of the *Alkimist,* whom he meant.

He was a great Peace-maker; if any of the neighbours fell out, he would never lett them alone till he had made them friends. He told a woman (his neighbour) that she laboured under the evill tongue of an ill neighbour (another woman) which came to her howse, who he sayd was a Witch.

He was sent Ambassador for Queen Elizabeth (Goody Faldo thinkes) into Poland. The Emperour of Muscovia, upon report of the great learning of the Mathematician, invited him to Mosco, with offer of two thousand pound a yeare, and from Prince Boris one thousand markes; to have his Provision from the Emperor's Table, to be honourably received, and accounted as one of the chief men in the Land. All of which Dee accepted not.

His regayning of the Plate for a friend's Butler, who comeing from London by water with a Basket of Plate, mistooke another basket that was like his. Mr J. Dee bid them goe by water such a day, and looke about, and he should see the man that had his basket, and he did so; but he would not gett the lost horses, though he was offered severall angells.

Arthur Dee, his sonne, a Physitian at Norwych and intimate friend of Sir Thomas Browne, M.D., told Dr Bathurst that (being but a Boy) he used to play at Quoits with the Plates of Gold made by Projection in the Garret of Dr Dee's Lodgings in Prague and that he had more than once seen the Philosopher's Stone.

(iii) from *The Alchemist* (1610) by Ben Jonson.

Sub.   Oh, I did look for him
With the sun's rising: 'marvel, he could sleep!
This is the day, I am to perfect for him
The magisterium, our great work, the stone;
And yield it, made, into his hands: of which,
He has this month talked, as he were possessed.
And now he's dealing pieces on't, away.
Methinks I see him, entering ordinaries,
Dispensing for the pox; and plaguey-houses,

Reaching his dose; walking Moorfields for lepers;
And offering citizens' wives pomander-bracelets,
As his preservative, made of the elixir;
Searching the spittle, to make old bawds young;
And the highways for beggars to make rich:
I see no end of his labours. He will make
Nature ashamed of her long sleep: when art,
Who's but a stepdame, shall do more than she,
In her best love to mankind, ever could.
If his dream last, he'll turn the age, to gold.   [I.iv]

    Sub.   It is, of the one part,
A humid exhalation, which we call
Materia liquida, or the unctuous water;
On the other part, a certain crass and viscous
Portion of earth; both which, concorporate,
Do make the elementary matter of gold:
Which is not, yet, *propria materia,*
But common to all metals, and all stones.
For where it is forsaken of that moisture,
And hath more dryness, it becomes a stone;
Where it retains more of the humid fatness,
It turns to sulphur, or to quick-silver:
Who are the parents of all other metals.
Nor can this remote matter suddenly
Progress so from extreme unto extreme,
As to grow gold, and leap o'er all the means.
Nature doth, first, beget the imperfect; then
Proceeds she to the perfect. Of that airy,
And oily water, mercury is engendered;
Sulphur o'the fat and earthy part: the one
(Which is the last) supplying the place of male,
The other of the female, in all metals.
Some do believe hermaphrodeity,
That both do act and suffer. But these two
Make the rest ductile, malleable, extensive.
And even in gold, they are; for we do find
Seeds of them, by our fire, and gold in them:
And can produce the species of each metal

More perfect thence than nature doth in earth.
Beside, who doth not see, in daily practice,
Art can beget bees, hornets, beetles, wasps,
Out of the carcasses and dung of creatures;
Yea, scorpions, of an herb, being ritely placed:
And these are living creatures, far more perfect
And excellent than metals.

    Sur.   What else are all your terms,
Whereon no one o' your writers grees with other?
Of your elixir, your *lac virginis,*
Your stone, your medicine, and your chrysosperm,
Your sal, your sulphur, and your mercury,
Your oil of height, your tree of life, your blood,
Your marcasite, your tutty, your magnesia,
Your toad, your crow, your dragon, and your panther,
Your sun, your moon, your firmament, your adrop,
Your lato, azoch, zernich, chibrit, heautarit,
And then, your red man, and your white woman,
With all your broths, your menstrues, and materials,
Of piss, and egg-shells, women's terms, man's blood,
Hair o' the head, burnt clouts, chalk, merds, and clay,
Powder of bones, scalings of iron, glass,
And worlds of other strange ingredients,
Would burst a man to name?
    Sub.   And all these, named
Intending but one thing: which art our writers
Used to obscure their art.
    Mam.   Sir, so I told him,
Because the simple idiot should not learn it,
And make it vulgar.
    Sub.   Was not all the knowledge
Of the Egyptians writ in mystic symbols?
Speak not the Scriptures oft in parables?
Are not the choicest fables of the Poets,
That were the fountains, and first springs of wisdom,
Wrapped in perplexed allegories?
                            [II.iii]

    Fac. He's busy with his spirits, but we'll upon him.

      .   .   .

Sub.   He first shall have a bell, that's Abel;
And, by it, standing one, whose name is Dee,
In a rug gown; there's D and Rug, that's Drug:
And, right anenst him, a Dog snarling Er;
There's Drugger, Abel Drugger. That's his sign.
And here's now mystery, and hieroglyphic!

                                       [II.vi]

(iv) from *The Tempest* (*c.*1611–12) by William Shakespeare.

Pro. Ye elves of hills, brooks, standing lakes, and groves;
And ye that on the sands with printless foot
Do chase the ebbing Neptune, and do fly him
When he comes back; you demi-puppets that
By moonshine do the green sour ringlets make,
Whereof the ewe not bites; and you whose pastime
Is to make midnight mushrooms, that rejoice
To hear the solemn curfew; by whose aid–
Weak masters though ye be–I have bedimm'd
The noontide sun, call'd forth the mutinous winds,
And 'twixt the green sea and the azur'd vault
Set roaring war. To the dread rattling thunder
Have I given fire, and rifted Jove's stout oak
With his own bolt; the strong-bas'd promontory
Have I made shake, and by the spurs pluck'd up
The pine and cedar. Graves at my command
Have wak'd their sleepers, op'd, and let 'em forth,
By my so potent art.

                                          [V.i]

Prospero Our revels now are ended. These our actors,
As I foretold you, were all spirits, and
Are melted into air, into thin air;
And, like the baseless fabric of this vision,
The cloud-capp'd towers, the gorgeous palaces,
The solemn temples, the great globe itself,
Yea, all which it inherit, shall dissolve,
And, like this insubstantial pageant faded,
Leave not a rack behind. We are such stuff
As dreams are made on; and our little life
Is rounded with a sleep.

                                          [IV.i]

APPENDIX A:
A NOTE ON
ENOCHIAN MAGIC

DEE's angel magic has been minimized by scholarly commentators, who choose instead to stress his concrete achievements, his breadth of mind, and the importance of his philosophy to the Elizabethan Renaissance. This emphasis is entirely proper: but Dee's magic cannot be and should not be excused away. The principal difficulty is that most of Dee's commentators do not understand it.

It should be clear that this angel magic must be very sharply distinguished from spiritualism, an activity in which individuals desperate for proof of an afterlife implore absolutely anything invisible to make contact with them. It is essential to understand what is meant by the term 'magic'. One can truthfully describe it as the practical application of Renaissance occult philosophy as described in my comment on Dee's *Letter To Sir William Cecil* and which is worth repeating here.

1. All is a Unity, created and sustained by God through His Laws.
2. These Laws are predicated upon Number.
3. There is an art of combining Hebrew letters and equating them with Number so as to perceive profound truths concerning the nature of God and His dealings with Man.
4. Man is of divine origin. Far from being created out of dust, as in the Genesis account, he is in essence a star daemon.
5. As such, he has come from God and must return to Him.
6. It is essential to regenerate the divine essence within Man, and this can be done by the powers of his divine intellect.
7. According to the Qabalah, God manifests by means of ten progressively more dense emanations: and Man, by dedicating his mind to the study of divine wisdom and by refining his whole being and by eventual communion with the angels themselves, may at last enter into the presence of God.
8. An accurate understanding of natural processes, visible and invisible, enables Man to manipulate these processes through the powers of his will, intellect and imagination.

9. The Universe is an ordered pattern of correspondence: or as Dee puts it: 'Whatever is in the Universe possesses order, agreement and similar form with something else.'[1]

    The aim of magic, then, is to make Man conscious of his own divinity so that he may progress in a straight line vertically to God, howsoever God may be conceived. In the course of doing magic, the aspirant may believe that he is encountering non-human beings. Magicians still differ as to whether these beings have objective or subjective existence.

    The best accounts of the subjectivist position are: Crowley's *The Initiated Interpretation of Ceremonial Magic,*[2] in which the spirits are equated with portions of the human brain; and Regardie's *The Art and Meaning of Magic,*[3] where spirits are equated with Freud's notion of complexes, the astral plane is explained as Jung's notion of the Collective Unconscious, and the gods invoked by the magician are shown to be Jungian archetypes. Although a number subscribe to this eminently sane position, and although the arguments of both Crowley and Regardie are cogent and persuasive, it should be noted that each one, for his own reasons, later recanted and embraced the objectivist view,[4] which was of course that of Dee.

    This position is simply that there is much more to reality than the physical universe of the materialist, that there is intelligent life in other dimensions, and that human beings may grow wiser and greater by means of their encounters with it.

    If the objectivist magician wishes to prove his case, he must show that there are praeter-human communications which display knowledge beyond the ken of the recipient. It is a further advantage if this knowledge can be fruitfully applied. In a purely practical sense, it is of no importance whether the objectivist or subjectivist paradigm describes the actual facts.[5] One is reminded of the controversy in contemporary philosophy of science between Pragmatist instrumentalism and the Realism of Professor Popper, which controversy, fascinating though it is, does not affect the continuing course of scientific discovery. As Aleister Crowley states simply: 'By doing certain things, certain things happen.'[6]

    Dee's contribution to magic consisted of a great mass of raw material. The quality of this material was erratic but it is the cornerstone of 'Enochian Magic'. The word 'Enochian' is derived from the Hebrew 'Enoch', the literal meaning of which is 'to initiate'. Part of Dee's magical legacy is the Angelic or Enochian language. Here is the essence of the matter.

1. In a way which is still not quite clear, Dee and Kelley obtained 100 squares filled with letters and usually numbering 49 × 49.[7]
2. Dee would have one or more of these squares before him.
3. Kelley sat at the Holy Table made according to angelic instruction and gazed at the shewstone, after a time seeing an angel who would point

    with a rod to letters in succession on one of these charts.

4.   Kelley would report, for instance: 'He points to column 5, rank 23', apparently not mentioning the letter, which Dee found and wrote down from the square before him.

5.   This implies that Kelley had absolutely no idea which words would be formed. To execute that feat, the man commonly denounced as a confidence trickster would have to have known the exact positions of the 2,401 letters in each of the tables. There must be an easier way of getting a living.

6.   Angels dictated the words backwards, warning that undesired forces could be evoked by pronouncing them the right way.

7.   Dee and Kelley rewrote the words forwards and the result was the Enochian Keys or Calls.

8.   There are nineteen. The first two conjure the Element called Spirit; the next sixteen invoke the Four Elements, each sub-divided into four; and the nineteenth, by changing two names, can be used to invoke any one of Thirty 'Aethyrs', 'Aires', or dimensions of existence.

9.   The language in which these Keys are written possesses a vocabulary; grammar, and syntax of its own.

10. All of which leaves sceptics and subjectivists with a genuine and interesting intellectual problem:

    a) 'Enochian' bears little relation to any known language.

    b) Yet philologists agree that it is impossible for a man to invent a new language.

       There are other issues too. The beauty of the Enochian Keys is apparent in English translation. As Crowley writes: 'To condemn Kelley as a cheating charlatan–the accepted view–is simply stupid. If he invented Enochian and composed this superb prose, he was at worst a Chatterton with fifty times that poet's ingenuity and five hundred times his poetical genius.' Moreover: 'The genuineness of these Keys is guaranteed by the fact that anyone with the smallest capacity for Magick finds that they work.[8] In Thomas Head's judgement; '. . . the most substantial and convincing proof of the *essential genuineness of both Dee and Kelley is their monumental ignorance of what to do with the material they have accumulated.'*[9]

       However, there were others who explored the paths trodden by Dee and Kelley. In the seventeenth century, Elias Ashmole appears to have tried Dee's system 1671–6.[10] One would like to know if anyone kept the tradition alive during the eighteenth century.[11] The pursuit of occult philosophy and practice in the nineteenth century by men such as Francis Barrett, Frederick Hockley, Kenneth MacKenzie, Robert Wentworth Little, and Sir Edward Bulwer-Lytton is a subject begging for scholarly exhumation. In the later years of Victoria's reign, however, we know for a fact that Dee's magical work became the crowning synthesis of the system studied and practised by the men and women members of the Hermetic Order of the Golden Dawn.

This Order was founded on the basis of a set of cipher manuscripts which came into the possession of Dr W. Wynn Westcott, a London coroner, in 1887. Westcott asked an occult scholar, S. L. 'MacGregor' Mathers, to assist him, and curiously enough they found that the code was contained in *Polygraphiae* by John Trithemius, whose *Steganographiae* had been so extolled by John Dee in his *Letter To Sir William Cecil*. The manuscripts contained skeletonic rituals of a loosely Rosicrucian nature and the address of one Fraulein Sprengel in Nuremberg. Westcott claimed that he wrote to her, receiving in return a Charter to found the Golden Dawn.

It has been alleged that Sprengel never existed and that Westcott was a fantasizing forger. [12] The controversy surrounding the Order's origins is not germane here. The facts remain that Mathers expanded and wrote up the skeletonic rituals and these were duly enacted in Temples set up in London, Edinburgh, Bradford, Weston-super-Mare, and later, Paris.

In 1891, Westcott claimed that Sprengel had broken off all communication, urging the Golden Dawn leaders to form their own links with 'the Secret Chiefs', supposedly superhuman beings concerned with the spiritual welfare of mankind. In Paris in 1892 Mathers claimed to have established these links. [13] A second, inner, and 'Rosicrucian' Order was founded, The Red Rose and the Cross of Gold, and page after page of occult teaching flowed from the clairvoyantly inspired pen of Mathers. Interestingly enough, the resulting system was beautiful and possessed bewildering yet logically coherent complexity.

It was this which caused the unjustly neglected and much under-valued author Arthur Machen to deplore the Golden Dawn system as being without true redeeming value. He wrote: '[The Golden Dawn] embraced all mythologies and all mysteries of all races and ages, and *referred* or *attributed* them to each other and proved that they all came to much the same thing; and that was enough! That was not the ancient frame of mind; it was not even the 1809 frame of mind. But it was very much the eighteen-eighty and later frame of mind.' [14]

All who love good literature must long for a sound revaluation of Arthur Machen's works which will make clear his excellence as a writer; but here he has missed the point entirely. It does not appear to have struck him that whilst 'it was very much the eighteen-eighty and later frame of mind', it was also quintessentially the frame of mind of the Renaissance. The beliefs and aims of the Golden Dawn magicians were those of Renaissance magi like Dee.

We are now in a better position to summarise the achievements of that strange man Mathers in the creation of the Golden Dawn system.

1. Mathers welded together Renaissance occult philosophy, including and especially the Qabalah, with certain of its sources which had come to light by his time–and his own inspiration.

2. The result was a body of knowledge and a method for taking practical advantage of that knowledge.

3. The entire system, the first nine volumes of which fill 870 pages in the latest edition, is summarized and synthesized again in more concentrated form within a refined paradigm deriving directly from Dee's angel magic.

4. The 'Adepts' who had mastered all the earlier knowledge and praxis consequently found themselves confronted by a new learning which incorporated and surpassed the old; providing the aspirant with new maps for the exploration of other dimensions of existence, methods for so doing, and a language for communication with beings thereby encountered, all of which was based on the writings of Dee.

However, the Golden Dawn curriculum demanded so much work when properly pursued that few even reached the stage of elementary grappling with the Enochian system until a member turned his attention to the matter: this was Aleister Crowley.

Most people who disparage Crowley have not read him. [15] It is possible that his reputation may remain blackened for as many centuries as that of Dee, with whom he shared a number of characteristics in common. Like Dee, Crowley was a man of many talents: a poet, mountaineer, explorer, big game hunter, yogi, essayist, novelist, chess master, and magus. A fundamental principle of Crowley's faith was: 'Every man and every woman is a star': [16] Dee shared the faith of Renaissance magi in Man as a star daemon. Dee sought for communion with 'angels', Crowley with 'praeter-human intelligences'. Despite this, Crowley identified with Kelley, believing him to have been a previous incarnation, and finding Dee 'unsupportable at times, with his pity, pedantry, credulity, respectability and lack of humour.' [17]

Crowley commenced his investigations of the Thirty Aethyrs in Mexico in 1900. The visions of the Thirtieth and Twenty-Ninth Aethyrs, recorded on 14 and 17 November 1900, are not especially satisfactory and he found himself unable to progress further until nine years later, when all the Aethyrs from the Twenty-Eighth to the First were explored in turn. An example forms the subject-matter of Appendix B.

Although Crowley's results have yet to be surpassed, much work was subsequently done on Enochian Magic by his one-time disciple, Dr Israel Regardie, member and Adept of a Golden Dawn offshoot in the 1930s. Regardie earned the lasting gratitude of occult students by breaking his oath of secrecy and publishing the Golden Dawn teachings, though not for financial gain. He defended his action ably,[18] arguing that the Order had become moribund, neglecting practical work in general and the Enochian system in particular. Shortly before his death he completed *The Complete Golden Dawn System of Magic,* Volume Ten of which contains the Enochian system, expert comment by colleagues, the fruits of Regardie's own labours on the matter, and his Enochian-English, English-Enochian dictionary.

In spite of schism, quarrels and much undignified squabbling, the Golden Dawn has survived into our own time and there are currently Temples in Atlanta, Georgia; Las Vegas, Nevada; Phoenix, Arizona; Los Angeles and San Diego, California; also Mexico City; and there may be others quietly pursuing their work. Much more practical work on Enochian Magic needs to be done by those sufficiently competent to tackle it. One hopes that those who succeed bear in mind the honoured place in the history of magic earned honestly by the pioneering studies of John Dee.

APPENDIX B:
ENOCHIAN MAGIC:
AN ITEM FOR
COMPARISON

IN November 1909, Aleister Crowley and his disciple Victor Neuberg were in the Algerian part of the Sahara, where they practised Enochian Magic by invoking in succession the Thirty Aethyrs or dimensions with their respective angels. The resulting records, known as *The Vision and the Voice,* were published in *The Equinox,* volume I, number v (London, 1911).

A number of commentators have demonstrated their inability to comment on this document, most notably John Symonds in *The Great Beast,* a biography of Crowley which is poisoned by hostility and riddled with factual inaccuracies. One is reminded of Dr Thomas Smith writing on John Dee. Symonds concentrates entirely on the sensationalism of the 10th Aethyr, in which an encounter with the demon Choronzon is described. The other visions are ignored. One is best advised to ignore Symonds and turn instead to the relevant chapter in Israel Regardie's study, *The Eye In The Triangle.*

Better still, one could turn to the original. Here is an example, *The Cry of the 22nd Aethyr.* Crowley is the seer and 'O.V.' refers to Neuberg, the scribe. The curious reader is recommended to compare the following with the records of John Dee.

The Cry of the 22nd Aethyr, which is called LIN
There comes first into the stone the mysterious table of forty-nine squares. It is surrounded by an innumerable company of angels; these angels are of all kinds, some brilliant and flashing as gods, down to elemental creatures. The light comes and goes on the tablet; and now it is steady, and I perceive that each letter of the tablet is composed of forty-nine other letters in a language which looks like that of Honorius; but when I would read, the letter that I look at becomes indistinct at once.

And now there comes an Angel, to hide the tablet with his mighty wing. This Angel has all the colours mingled in his dress; his head is proud and beautiful; his headress is of silver and red and blue and gold and black, like cascades of water, and in his left hand he has a pan-pipe of the seven holy metals, upon which he plays. I cannot tell you how wonderful the music is, but it is so wonderful that one only lives in one's ears; one cannot see anything any more.

Now he stops playing and moves with his finger in the air. His finger leaves a trail of five of every colour, so that the whole Aire is become like a web of mingled lights. But through it all drops dew.

(I can't describe these things at all. Dew doesn't represent what I mean in the least. For instance, these drops of dew are enormous globes, shining like the full moon, only perfectly transparent, as well as perfectly luminous.)

And now he shows the tablet again, and he says: As there are 49 letters in the tablet, so are there 49 kinds of cosmos in every thought of God. And there are 49 interpretations of every cosmos and each interpretation is manifested in 49 ways. Thus also are the calls 49, but to each call there are 49 visions. And each vision is composed of 49 elements, except in the 10th Aethyr, that is accursed, and that hath 42.

All this while the dewdrops have turned into cascades of gold finer than the eyelashes of a little child. And though the extent of the Aethyr is so enormous, one perceives each hair separately as well as the whole thing at once. And now there is a mighty concourse of angels rushing toward me from every side, and they melt upon the surface of the egg in which I am standing in the form of the god Kneph, so that the surface of the egg is all one dazzling blaze of liquid light.

Now I move up against the tablet, I cannot tell you with what rapture. And all the names of God that are not known even to the angels, clothe me about.

All the seven senses are transmuted into one sense, and that sense is dissolved in itself . . . (Here occurs Samadhi) . . . let me speak, O God; let me declare it . . . all. It is useless; my heart faints, my breath stops. There is no link between me and P . . . I withdraw myself. I see the table again.

(He was behind the table for a very long time. *O.V.*)

And the table burns with intolerable light; there has been no such light in any of the Aethyrs until now. And now the table draws me back into itself; I am no more.

My arms were out the the form of a Cross, and that Cross was extended, blazing with light into infinity. I myself am the minutest point in it. This is *the birth of form.*

I am encircled by an immense sphere of many-coloured bands; it seems it is the sphere of the Sephiroth projected in three dimensions. This is *the birth of death.*

Now in the centre within me is a glowing sun. That is *the birth of hell.*

Now all is swept away, washed away by the table. It is the virtue of the table to sweep everything away. It is the letter I in this Aethyr that gives this vision, and: is its purity, and N is its energy. Now everything is confused, for I invoked the Mind, that is disruption. Every Adept who beholds this vision is corrupted by mind. Yet it is by virtue of mind that he endures it and passes on, if so be that he pass on. Yet there is nothing higher than this for it is perfectly balanced in itself. I cannot read a word of the holy Table, for the letters of the Table are all wrong. They are only the shadows of shadows. And whoso beholdeth this Table with this rapture, is light. The true word for light hath seven letters. They are the same as ARARITA, transmuted.

There is a voice in this Aethyr but it cannot be spoken. The only way one can represent it is as a ceaseless thundering of the word Amen. It is not a repetition of Amen, because there is no time. It is one Amen continuous. Shall mine eye fade before thy glory. I am the eye. That is why the eye is seventy. You can never understand why, except in this vision.

And now the table recedes from me. Far, far it goes, streaming with light. And there are two black angels bending over me, covering me with their wings, shutting me up into the darkness and I am lying in the Pastos of our Father Christian Rosenkreutz, beneath the Table in the Vault of seven sides. And I hear these words:

The voice of the Crowned Child, the Speech of the Babe that is hidden in the egg of blue. (Before me is the flaming Rosy Cross). I have opened mine eye, and the Universe is dissolved before me, for force is mine upper eye-lid and matter is my lower eye-lid. I gaze into the seven spaces, and there is naught.

The rest of it comes without words; and then again.

I have gone forth to war, and I have slain him that sat upon the sea, crowned with the winds. I put forth my power and he was broken. I withdrew my power and he was ground into dust. Rejoice with me, O ye Sons of the Morning; stand with me upon the Throne of Lotus; gather yourselves up unto me, and we shall play together in the fields of light. I have passed into the Kingdom of the West after my Father.

Behold! where are now the darkness and the terror and the lamentation? For ye are born into the new Aeon; ye shall not suffer death. Bind up your girdles of gold! Wreathe yourselves with garlands of my unfading flowers! In the nights we will dance together, and in the morning we will go forth to war; for, as my Father liveth that was dead, so do I live and shall never die.

And now the table comes rushing back. It covers the whole stone, but this time it pushes me before it, and a terrible voice cries: Begone! Thou has profaned the mystery; thou hast eaten of the shew-bread; thou hast spilt the consecrated wine! Begone! For the Voice is accomplished. Begone! For that which was open is shut. And thou shalt not avail to open it, saving by virtue of him whose name is one, whose spirit is one, whose individuum is one, and whose permutation is one; whose light is one, whose life is one, whose love is one. For though thou art joined to the inmost mystery of the heaven, thou must accomplish the sevenfold task of the earth, even as thou sawest the Angels from the greatest unto the least. And of all this shalt thou take back with thee but a little part, for the sense shall be darkened, and the shrine re-veiled. Yet know this for thy reproof, and for the stirring up of discontent in them whose swords are of lath, that in every word of this vision is concealed the key of many mysteries, even of being, and of knowledge, and of bliss; of will, of courage, of wisdom, and of silence, and of that which being all these, is greater than all these. Begone! For the night of life is fallen upon thee. And the veil of light hideth that which is.

With that, I suddenly see the world as it is, and I am very sorrowful.
BOU-SAADA.
November 28 1909, 4-6 p.m.
(Note: You do not come back in any way dazed; it is like going from one room into another. Regained normal consciousness completely and immediately.)

APPENDIX C:
WHAT IS QABALAH?

The following extract from Aleister Crowley's *Liber 777* (London, 1909) gives the most succinct and practical answer in print.

WHAT IS QABALAH?

Qabalah is

(a) A language fitted to describe certain classes of phenomena and to express certain classes of ideas which escape regular phraseology. You might as well object to the technical terminology of chemistry.

(b) An unsectarian and elastic terminology by means of which it is possible to equate the mental processes of people apparently diverse owing to the constraint imposed upon them by the peculiarities of their literary expression. You might as well object to a lexicon or a treatise on comparative religion.

(c) A system of symbolism which enables thinkers to formulate their ideas with complete precision and to find simple expression for complex thoughts, especially such as include previously disconnected orders of conception. You might as well object to algebraic symbols.

(d) An instrument for interpreting symbols whose meaning has become obscure, forgotten or misunderstood by establishing a necessary connection between the essence of forms, sounds, simple ideas (such as number) and their spiritual, moral or intellectual equivalents. You might as well object to interpreting ancient art by consideration of beauty as determined by physiological facts.

(e) A system of omniform ideas so as to enable the mind to increase its vocabulary of thoughts and facts through organising and correlating them. You might as well object to the mnemonic value of Arabic modifications of roots.

(f) An instrument for proceeding from the known to the unknown on similar principles to those of mathematics. You might as well object to the use of $\sqrt{\ }$, -1, x^4 etc.

(g) A system of criteria by which the truth of correspondence may be tested with a view to criticizing new discoveries in the light of their coherence with the whole body of truth. You might as well object to judging character and status by educational and social convention.

NOTES

INTRODUCTION

1. Dee, *The Compendious Rehearsal.*
2. ibid.
3. William Lilly, *The History of His Life and Times* (London, 1774).
4. John Aubrey, *The Natural History and Antiquities of the County of Surrey,* 5 vols. (London, 1718–19).
5. Meric Casaubon, ed. *A True & Faithful Relation of what passed for many Years Between Dr: John Dee . . . and Some Spirits* (London, 1659).
6. Thomas, Smith, *The Life of John Dee,* tr. William A. Ayton (London, 1908).
7. William Godwin, *Lives of the Necromancers* (London, 1834).
8. Charlotte Fell Smith, *John Dee: 1527-1608* (London, 1909).
9. Further confirmation of this view can be found in David W. Waters, *The Art of Navigation in England in Elizabethan and Early Stuart Times.*
10. F. R. Johnson, *Astronomical Thought in Renaissance England* (Baltimore, 1937).
11. For example, *The Occult Philosophy In The Elizabethan Age* (London, 1979). (But see Bibliography.)
12. Peter French, *John Dee: The World of an Elizabethan Magus* (London, 1972).
13. Wayne Shumaker, ed. *John Dee on Astronomy* (University of California Press: Berkeley, 1976).
14. Richard Deacon, *John Dee: Scientist, Geographer, Astrologer and Secret Agent* (London, 1968).
15. Yates, *The Occult Philosophy.*
16. ibid.
17. See Appendix A.

SUPPLICATION TO QUEEN MARY

1. French, *John Dee.*
2. Dee, *Compendious Rehearsal.*

3. Jayne Sears, *Library Catalogues of the English Renaissance* (Berkeley and Los Angeles, 1956).
4. Frances Yates, *Theatre of the World.* (London and Chicago, 1969).
5. It is difficult to render monetary sums of the past in terms of the present. Caution is necessary: but I concur with the suggestion of Eric Towers, an historian of the eighteenth century aristocracy, that roughly accurate calculations can be made in terms of the price of a loaf of bread.
6. Dee is of course referring to the dissolution of the monasteries and the plundering of religious houses under Northumberland.

PROPAEDEUMATA APHORISTICA

1. Dee, *Preface to Euclid.*
2. University of California Press, Berkeley, 1978.
3. This was, of course, the renowned cartographer, whom Dee met in the Low Countries in 1547. Dee brought two Mercator globes back to England.
4. The branch of physics that deals with heat.
5. French, *John Dee.*
6. *John Dee's Astronomy,* tr. Shumaker.

TO SIR WILLIAM CECIL

1. Yates, *The Occult Philosophy In The Elizabethan Age.*
2. ibid.
3. ibid.
4. quoted in ibid.
5. ibid.
6. French, *John Dee.*
7. Yates, *The Occult Philosophy.*
8. Bailey, *Notes and Queries* (31 May 1879).
9. D. P. Walker, *Spiritual and Demonic Magic: From Ficino to Campanella* (Warburg Institute, University of London, 1958).
10. Isaac D'Israeli, *The Amenities of Literature,* ed. Earl of Beaconsfield (London, n.d.).
11. Bailey, *Notes And Queries.*
12. Quoted, Bailey, *Notes and Queries.*

MONAS HIEROGLYPHICA

1. *The Hieroglyphic Monad,* tr. J. W. Hamilton-Jones (London, 1947); New York 1975, 1977, with Preface by Diane di Prima.
2. *Monas Hieroglyphica,* tr. C. H. Josten, *Journal of the Society for the Study of Alchemy and Early Chemistry,* XII (1964).

3. Yates, *The Occult Philosophy.*
4. French, *John Dee.*
5. One could also compare the 'Art' of Ramon Lull, the 'Key' of William Postel, the Tarot and Crowley's Naples Arrangement as applied to the Cabalistic Tree of Life.
6. Letter to Maximilian, tr. Josten.
7. One is irresistibly reminded of Dee's *Letter To Sir William Cecil* in which he wrote of 'men hard to find, although daily seen'. This passage is taken from the dedicatory letter to Maximilian, tr. Josten.
8. By Christian Rosencreutz (pseudonym of Johann Valentin Andreae), tr. E. Foxcroft (London, 1690).
9. One presumes that by this word Dee means 'physical'.
10. Compare the last sentence of the extract from Theorem XX: '. . . and because I have already terminated my discourse for those whose *gaze is centered within the heart,* it is now necessary to translate my words for those *whose heart is centered within their eyes.*' (Italics mine.) Students of Yoga will think of the *chakras,* those centres of power and wisdom within the body which correspond with but are not to be identified with the endocrine glands. A Master of Yoga is one who has awakened 'the Third Eye' or Ajna chakra, between and behind the eyebrows and the repository of spiritual insight and divine wisdom rightly exercised.
11. di Prima, *Preface.*

PREFACE TO EUCLID

1. *John Dee's Astronomy.*
2. *Mathematical Preface* to the *Elements of Geometry,* ed. Thomas Rudd (London, 1651) and *Mathematical Preface* to *Euclid's Elements of Geometry,* ed. John Leeke and George Serle (London, 1661).
3. *The Occult Philosophy.*
4. E. G. R. Taylor, *Tudor Geography: 1485-1583,* Appendix 8 (London, 1930).

GENERAL AND RARE MEMORIALS

1. E. G. R. Taylor, *Tudor Geography: 1485-1583* (London, 1930).
2. ibid.
3. Appendix 8B, ibid.
4. ibid. The departure table, of about 1556, survives in manuscript (Ashm. 242, No. 43 Bodleian) and has been described in a paper by Professor E. G. R. Taylor ('John Dee and the Nautical Triangle, 1575', *Journal of Inst. of Navigation,* vol. 8).
5. Yates, *The Occult Philosophy.*
6. Taylor, *Tudor Geography.*

DIARIES 1577–83

1. Yates, *The Occult Philosophy.*
2. French, *John Dee.*
3. *Letter Book A.D. 1573-1580,* ed. E. J. L. Scott, Camden Society Publications, XXXIII (London, 1884).
4. Yates, *The Occult Philosophy.*
5. Hippocrates Junior (pseud.) (ed.), *The Predicted Plague* (London, 1889).
6. James O. Halliwell (ed.), *The Private Diary,* Camden Society Publications, XIX (London, 1842).

SPIRITUAL DIARIES 1583–7

1. See *Diaries,* 9 March 1582 in previous chapter.
2. See *An Introduction To The Enochian Teaching and Praxis* by Thomas Head, Ph.D. (Oxon) in *The Complete Golden Dawn System of Magic,* Volume Ten (Phoenix, Arizona, 1984). The entire work is edited and commented upon by Israel Regardie and Head's essay is warmly recommended.
3. Dee used a number of shewstones. A crystal and his 'magic mirror' can be seen in the British Museum. The latter, which is black and made of obsidian, was obtained from the Aztecs by a member of Cortes' Mexican expedition. Long after Dee's death, it passed into the hands of Horace Walpole who glued beneath it the sneer of Samuel Butler in *Hudibras* (1664):

> Kelley did all his Feats upon
> *The Devil's Looking-glass,* a stone
> Where playing with him at *Bo-peep*
> He solved all problems ne'er so deep.

The history of the stone is explored by Hugh Tait in ' "The Devil's Looking-Glass": The Magical Speculum of Dr. John Dee', in *Horace Walpole: Writer, Politician and Connoisseur,* ed. Warren Hunting Smith (New Haven and London, 1967). The British Museum acquired the object in October 1966.
4. Head, *An Introduction To The Enochian Teaching And Praxis.*
5. ibid.
6. The latter is a large, beautifully bound photofacsimile of the original and is introduced by Stephen Skinner, a noted contemporary scholar of Enochian.
7. Subsequently, Madini becomes Madimi.
8. *Introduction* to the London 1976 edition of Casaubon.
9. Introduction to *The Hieroglyphic Monad.*
10. Charlotte Fell Smith, *John Dee 1527-1608* (London, 1909).

11. Phoenix, Arizona, 1982.
12. Yates, *The Occult Philosophy*.
13. Head, *Introduction*.
14. French, *John Dee*.
15. Head, *Introduction*.
16. Donald C. Laycock, *The Complete Enochian Dictionary* (London, 1978).
17. *Hamlet*, IV.

LETTER TO QUEEN ELIZABETH

1. Shumaker, *John Dee's Astronomy*.
2. French, *John Dee*.
3. Yates, *The Occult Philosophy*.

PERSONAL DIARIES 1589–95

1. Smith, *John Dee*.
2. Walter I. Trattner, 'God and Expansion in Elizabethan England: John Dee, 1527–1608, *Journal of the History of Ideas*, XXV (1964).
3. C. H. Josten, 'An Unknown Chapter in the Life of John Dee', *Journal of the Warburg and Courtauld Institutes*, XXVIII (1965).
4. French, *John Dee*.
5. French, *John Dee*. The author quotes John Foxe's *Actes and Monuments* (1563), p. 1414.
6. John Aubrey, *Brief Lives*. See the chapter headed 'Some Opinions' for the full extract.
7. ibid.
8. *The Predicted Plague* (London, 1889).
9. Aubrey, *Brief Lives*.

LETTER TO THE ARCHBISHOP OF CANTERBURY 1599

1. It is instructive to compare the attitude of Sunni or 'mainstream' Muslims to the Sufis, mystics who profess devotion to Islam but who are regarded with suspicion by the orthodox.
2. Yates, *The Occult Philosophy*.

PERSONAL DIARIES 1600–1

1. Yates, *The Occult Philosophy*.
2. French, *John Dee*, ch. 8, 'John Dee as an Antiquarian'.

LETTER TO KING JAMES I, 1604

1. Russell Hope Robbins, F.R.C.L., *An Encyclopaedia of Witchcraft and Demonology* (London, 1964).

2. Dee also petitioned *To The Honorable Assemblie of the COMMONS in the present Parlament* (London, 1604), making his appeal in verse. This was not a wise decision as the first verse makes abundantly clear:

> To honour due unto you all
> And reverence to you each one
> I do first yield most special:
> Grant me this time, to hear my moane.

One has to agree with Peter French that Dee's poetry is 'abominable'.

A LATER SPIRITUAL DIARY

1. Dee's continuing pursuit of practical magic is further demonstrated by a manual of invocation, written in his own hand: *Tuba Veneris (c.1600) Libellus Veneri nigro Dee,* in the Yorke Collection of the Warburg Institute.
2. Yates, *The Rosicrucian Enlightenment* (London, 1972).
3. Yates, *The Occult Philosophy.*

SOME OPINIONS OF DEE 1600–1700

1. Deacon, *John Dee.* 'Richard Deacon' is the pseudonym of Donald McCormack, author of books on, *inter alia,* Jack the Ripper and the Hell-Fire Club, so-called.
2. Yates, *The Occult Philosophy.*
3. ibid. See also Yates, *Shakespeare's Last Plays: A New Approach* (London, 1975).

APPENDIX A

1. *Aphorisms.*
2. The essay is a footnote to *The Sword or Song; The Collected Works of Aleister Crowley,* Volume II.
3. Sangraal Foundation, Texas, 1969.
4. Crowley, for instance, in *MAGICK In Theory and Practice* and *Magick Without Tears;* Regardie most notably in *The Complete Golden Dawn System of Magic* and also in private conversations with the present editor.
5. 'The world is all the facts that are the case', Ludwig Wittgenstein, *Tractatus Logico-Philosophicus,* Proposition 1i.
6. *Liber O* in *MAGICK.*
7. Compare the use of 49 in Appendix B.
8. *The Confessions of Aleister Crowley.*
9. Head's *Introduction, The Complete Golden Dawn System,* Vol. 10.
10. Bodleian Ashmole MS 1790, art. 3.
11. Does the answer lie in the library of Sir Francis Dashwood? He is another figure whose reputation has suffered from vilification and who demands

further investigation. Fortunately, a biography by Eric Towers, shedding new light on the matter, is now available (*Dashwood: The Man and the Myth,* Crucible, 1986).

12.   The origins controversy can be studied in Ellic Howe's *The Magicians of the Golden Dawn* (London, 1972) in which it is skilfully argued that the Order was founded on a fraud, and my *Suster's Answer to Howe* (in Regardie's *What You Should Know About the Golden Dawn,* Phoenix, 1983), which raises questions Howe has failed to consider and arrives at a verdict of Not Proven.

        A third possibility has since been suggested in private conversation by Eric Towers: that Westcott was a forger but that the result was nevertheless enough for Mathers to bring through genuine Magic. This possibility should be seriously considered.

        In the end, of course, this question of origins is of purely academic interest. Either Golden Dawn magic works or it doesn't.

13.   *Mathers' Manifesto* in *What You Should Know About The Golden Dawn.*

14.   Machen, *Autobiography* (London, 1923).

15.   The best known but least interesting facts about Crowley are that he liked sex and took drugs; but so do so many people.

16.   *The Book of the Law,* I 3.

17.   Crowley, *Confessions.*

18.   *What You Should Know About the Golden Dawn.*

SELECT BIBLIOGRAPHY

A. MANUSCRIPTS BY DEE

The best bibliography of Dee available is that of Peter French in *John Dee,* which includes Dee's manuscript works.

B. PRINTED WORKS BY DEE

Autobiographical Tracts of Dr John Dee, warden of the College of Manchester, ed. James Crossley, *Chetham Society Publications* vol. XXIV (Manchester, 1851).

Diary for the years 1595-1601, ed. John E. Bailey (privately printed, 1880).

Diary, ed. Hippocrates Junior (pseud.) in *The Predicted Plague* (London, 1889). [According to French, 'the "Astrology of her Most Sacred and Illustrious Majestie Queene Elizabeth of Armada Renowne" assigned to Dee is spurious.']

General and Rare Memorials pertayning to the Perfect Arte of Navigation (London, 1577).

A Letter, Containing a Most Briefe Discourse Apologeticall with a Plaine Demonstration, and Fervent Protestation, for the Lawfull, Sincere, Very Faithfull and Christian Course of the Philosophical Studies and Exercises, of a Certaine Studious Gentleman (London, 1599). [To the Archbishop of Canterbury.]

Letter To Sir William Cecil, ed. John E. Bailey, *Notes and Queries,* 5th Series, XI (May 1879).

Mathematicall Preface to *The Elements of Geometrie of the most auncient Philosopher Euclide of Megara,* tr. Sir Henry Billingsley, ed. John Dee (London, 1570).

Monas Hieroglyphica, tr. C. H. Josten, *Journal of the Society for the study of Alchemy and Early Chemistry* XII (1964).

The Hieroglyphic Monad, tr. J. W. Hamilton-Jones (New York, 1977).

The Petty Navy Royal, ed. E. Arber, in *An English Garner,* vol. 2 (London, 1879).

The Private Diary, ed. James O. Halliwell, *Camden Society Publications,* vol. XIX (London, 1842).

To the Honorable Assemblie of the Commons in the present Parlament (London, 1604).

To the King's most excellent Majestie (London, 1604).

A True & Faithful Relation of what passed for many Yeers Between Dr: John Dee ... and Some Spirits, ed. Meric Casaubon (London, 1659; London, 1976).

An Unknown Chapter in the Life of John Dee, ed. C. H. Josten.
Journal of the Warburg and Courtauld Institutes, XVIII (1965).

C. OTHER SOURCES

Agrippa, Henry Cornelius, *Three Books of Occult Philosophy,* tr. James French (London, 1651).

Ainsworth, William Harrison, *Guy Fawkes, or The Gunpowder Treason* (London, n.d.).

Ashmole, Elias, *His Autobiographical and Historical Notes his Correspondence, and other Contemporary Sources Relating to his Life and Work,* ed. C. H. Josten, 5 vols. (Oxford, 1966).

Aubrey, John, *The Natural History and Antiquities of the County of Surrey,* 5 vols. (London, 1718–19).

——, *Brief Lives,* ed. Oliver Lawson Dick (London, 1958).

Besterman, Theodore, *Crystalgazing: A Study in the History, Distribution, Theory and Practice of Skrying* (London, 1924).

Butler, E. M., *The Myth of the Magus* (Cambridge, 1948).

——, *Ritual Magic* (Cambridge, 1949).

——, *The Fortunes of Faust* (Cambridge, 1952).

Carre, Meyrick H., 'Visitors to Mortlake: The Life and Misfortunes of John Dee, in *History Today,* XII (1962).

Clulee, Nicholas H., 'Astrology, Magic and Optics: Facets of John Dee's Early Natural Philosophy', in *Renaissance Quarterly,* XXX, 4 (Winter, 1977) [Renaissance Society of America].

Crowley, Aleister, *Liber 777* (London, 1909).

——, 'The Vision and the Voice', in *The Equinox,* I, V (London, 1911).

——, *Magick Without Tears* (Phoenix, 1982).

Dalton, Ormonde M., 'Notes on Wax Discs used by Dr Dee', in *Proceedings of the Society of Antiquaries of London,* XXI (1906–7).

Deacon, Richard, *John Dee: Scientist, Geographer, Astrologer and Secret Agent to Elizabeth I* (London, 1968).

D'Israeli, Isaac, *The Amenities of Literature,* ed. Earl of Beaconsfield (London, n.d.)

Godwin, William, *Lives of the Necromancers* (London, 1834).

Hort, Gertrude M., *Dr John Dee: Elizabethan Mystic and Astrologer* (London, 1922).

James, Montague R., 'Manuscripts formerly owned by Dr John Dee with Preface and Identifications', in *Supplement to the Bibliographical Society's Transactions* (London, 1921).

Jayne, Sears, *Library Catalogues of the English Renaissance* (Berkeley and Los Angeles, 1956).

Johnson, Francis R., *Astronomical Thought in Renaissance England* (Baltimore, 1937).

Kippis, Andrew, *Biographia Britannica,* 5 vols. (London, 1778–93).

Laycock, Donald C., *The Complete Enochian Dictionary* (London, 1978).

Lilly, William, *The History of His Life and Times* (London, 1774).

Read, Conyers, *Mr Secretary Walsingham and the Policy of Queen Elizabeth* (Oxford, 1925).

Regardie, Israel, *The Complete Golden Dawn System of Magic* (Phoenix, 1984).

Scholem, G. G., *Major Trends in Jewish Mysticism* (Jerusalem, 1941).

Shumaker, Wayne, *John Dee's Astronomy* (Berkeley, 1978).

Smith, Charlotte Fell, *John Dee: 1527-1608* (London, 1909).

Smith, Thomas, *The Life of John Dee*, tr. W. A. Ayton (London, 1908).

Spenser, Edmund. *Poetical Works*, ed. James C. Smith and Ernest De Selincourt (London, 1966).

Szonyi, T., 'Dee and Central Europe,' in *The Hungarian Studies in English*, XII.

Tait, Hugh, 'The Devil's Looking Glass: The Magical Speculum of Dr. John Dee', in *Horace Walpole: Writer, Politician and Connoisseur*, ed. Warren Hunting Smith (New Haven and London, 1967).

Taylor, E. G. R., *Tudor Geography: 1485-1583* (London, 1930).

Trattner, Walter I., 'God and Expansion in Elizabethan England: John Dee, 1527–1608', in *Journal of the History of Ideas*, XXV (1964).

Walker, D. P., *Spiritual and Demonic Magic from Ficino to Campanella* (London, 1958).

Waters, David W., *The Art of Navigation in England in Elizabethan and Early Stuart Times* (London, 1958).

Yates, Frances A., *The Art of Memory* (London and Chicago, 1966).

——, *Giordano Bruno and the Hermetic Tradition* (London and Chicago, 1964).

——, *Theatre of the World* (London and Chicago, 1969).

——, *Shakespeare's Last Plays: A New Approach* (London, 1975).

——, *The Rosicrucian Enlightenment* (London, 1972).

——, *The Occult Philosophy in the Elizabethan Age* (London, 1979).

Yeandle, W. H., *The Quadricentennial of the Birth of Dr John Dee* (London, 1927).

D. A NOTE ON BIOGRAPHY

Although the work of Thomas Smith remained the standard biography until 1909, and is worth a perusal for anyone who delights in the inability of a small mind to understand a great one, it is nevertheless a bad book. The biography of Charlotte Fell Smith is well-intentioned and still of value for the account of Dee's Continental period 1583-9. Richard Deacon's book raises important questions. The works of Frances Yates are essential for a proper understanding of Dee. The best biography to date is Peter French's *John Dee: The World of an Elizabethan Magus* (London, Boston and Henley, 1972). It should be noted, however, that French does not fully explore Dee's angel magic and acknowledges that there are many problems requiring further research if a true comprehension of Dee is to be attained.

Of further interest . . .

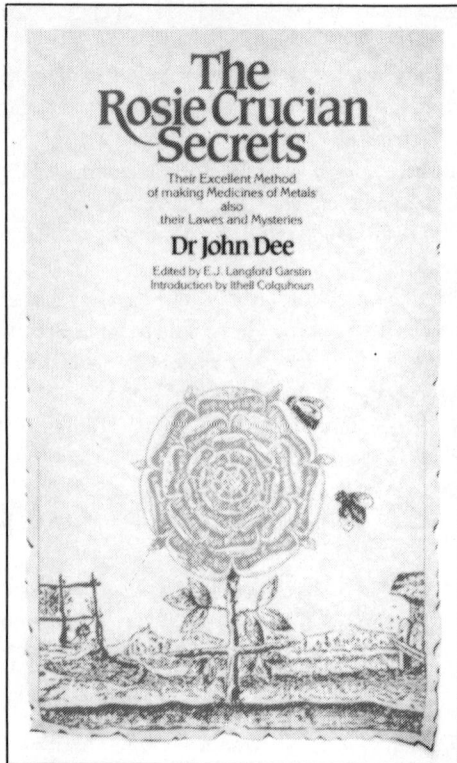

THE ROSIE CRUCIAN SECRETS

Timely reproduction of the last surviving copy of **John Dee's** *original work.*
Thoroughly absorbing and thought-provoking, it points to the existence
of a secret Elizabethan fraternity in England, and possibly across the
channel, to which Dee belonged, and among the members of which
circulated certain private manuscripts. *The most reasonable conclusion is
that this was a specialised branch of the Rosicrucian order.* Subjects discussed
include: the operation of making pearls; preparation of the great
philosopher's stone; making an ounce of gold out of half an ounce;
making the salt of gold and, probably the most contentious, *the laws
and mysteries of the Rosie Crucians.* This important and fascinating hermetic
text is here published for the first time, complete with full explanatory
and critical notes.

THE HEPTARCHIA MYSTICA OF JOHN DEE

A Primer of Hermetic Science and Magical Procedures by the Elizabethan Scholar Image

Robert Turner here presents a rare chance for students of the occult to gain access to the writings of a man hailed in the sixteenth century as the most learned man in Europe. Even today scholars still acknowledge the importance of his contributions in the areas of mathematics, astronomy and geography. The Heptarchia Mystica is Dee's basic guide to a system of occultism orientated strongly towards practical experimentation. Robert Turner has here painstakingly collated two of Dee's original manuscripts, at present held by the British library, supplemented them with his own notes, explanations and diagrams, and produced a valuable, and simplified, contribution to occultism.

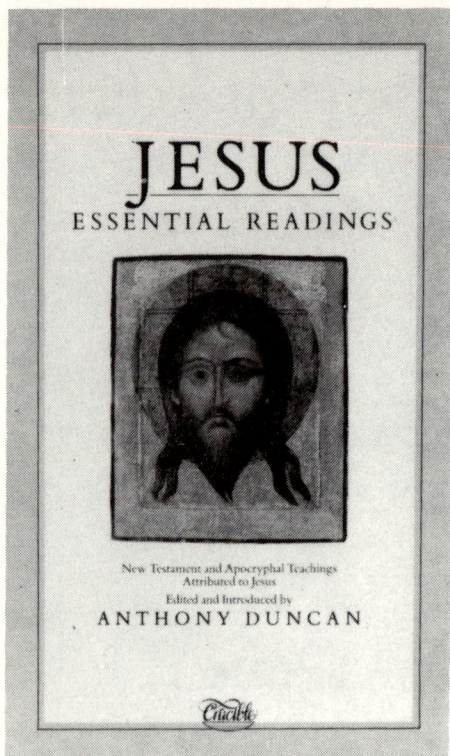

JESUS

Essential Readings

Anthony Duncan. At last — a chance for the lay person to study the teachings of Jesus without the usual biblicism and ecclesiasticism. Inspirationally translated into modern prose and verse — poetry was almost certainly its original form — this is a challenging and enlightening presentation. *An encounter with Jesus which is both unexpected and refreshing.* Aided by an introduction outlining the development of the four gospels and the sources from which they were compiled, the teachings themselves are thoughtfully grouped according to their theme. This fascinating yet easy-to-read volume ends beautifully and triumphantly with the 'Hymn of Jesus' popularly supposed to have been sung at the last supper.